CELL-CULTURE TEST METHODS

A symposium sponsored by
ASTM Committee F-4 on Medical and Surgical Materials and Devices
Dearborn, Mich., 11 May 1982

ASTM SPECIAL TECHNICAL PUBLICATION 810
S. A. Brown, University of California at Davis, editor

ASTM Publication Code Number (PCN)
04-810000-54

1916 Race Street, Philadelphia, Pa. 19103

Library of Congress Catalog Card Number: 83-70421

NOTE

The Society is not responsible, as a body,
for the statements and opinions
advanced in this publication.

Printed in Ann Arbor, Mich.
October 1983

Foreword

The symposium on Cell-Culture Test Methods was presented at Dearborn, Mich., 11 May 1982. The symposium was sponsored by ASTM Committee F-4 on Medical and Surgical Materials and Devices, S. A. Brown, University of California, Davis, Calif., presided as chairman and editor of the publication.

Related ASTM Publications

Medical Devices: Measurements, Quality Assurance and Standards, STP 810 (1983), 04-800000-54

Corrosion and Degradation of Implant Materials, STP 684 (1979), 04-684000-27

A Note of Appreciation to Reviewers

The quality of the papers that appear in this publication reflects not only the obvious efforts of the authors but also the unheralded, though essential, work of the reviewers. On behalf of ASTM we acknowledge with appreciation their dedication to high professional standards and their sacrifice of time and effort.

ASTM Committee on Publications

ASTM Editorial Staff

Janet R. Schroeder
Kathleen A. Greene
Rosemary Horstman
Helen M. Hoersch
Helen P. Mahy
Allan S. Kleinberg
Virginia M. Barishek
Susan L. Gebremedhin

Contents

Introduction

The ASTM Committee F-4 on Medical and Surgical Materials and Devices was organized in 1962 to develop standards on materials, test methods, and devices used in medicine and surgery. Since its inception, the committee has continued to address the question of the biological response of tissues to implanted materials and devices. While the actual definition of biocompatibility depends upon the nature of the implantation application, it has always been recognized that the biological response to a material plays a major role in the successful performance of the device made out of that material. Thus, the committee's efforts have focused on developing methods of testing for the biological response or biocompatibility of materials and incorporation of biocompatibility language in material standards.

During the early years of Committee F-4, the Biocompatibility Subcommittee was led and directed by Drs. Jonathan Cohen and Patrick Laing, both of whom were active clinicians and researchers in the field of orthopedics. The major product of their efforts was the development of a mutual trust and understanding between manufacturers, physicians, and academic researchers. This trust, which grew out of the educational interchange debate (heated at times) and familiarity with each others' problems, bore fruit in 1972 with the publication of the first standard for biocompatibility testing, ASTM Practice for Assessment of Compatibility of Metallic Materials on Tissue (F 361). This standard, which was based on the work of Drs. Cohen and Laing, outlines a comparative soft tissue and bone implant model, using various implantation times, control materials, and several animal species to assay for the biological response in situations similar, but not identical, to those encountered by orthopedic implants. Its flavor is orthopedic, as was the flavor of Committee F-4 at the time of its development. Six years later, in 1978, a second biocompatibility test method was approved by Committee F-4. ASTM Practice for Assessment of Compatibility of Nonporous Polymeric Materials for Surgical Implants with Regards to Effect of Materials on Tissue (F 469) was similar in scope and nature to ASTM F 361 with the exception that it was a method for assaying the long-term response of tissue to nonporous polymeric materials. Again, the nature of the method was related to long-term orthopedic applications.

In the mid-seventies, under the initiative of Drs. Jonathan Black and Stanley Brown, the activities of the biocompatibility section expanded beyond consideration of long-term implantation of materials in contact with muscle and bone, to encompass the questions of other types of materials and devices and the questions of testing of new materials and biological assays for screening materials and devices on a lot-to-lot basis. Our efforts were directed toward the question of how to test materials for biologic response when the intended end use applications were short- or long-term contact with different types of human tissues. In some instances we took documents from the United States Pharmacopia (USP) and modified them for application to medical devices. In 1979 ASTM Practice for Extraction of Medical Plastics (F 619) was published. In 1981 two short-term in vivo screening methods were approved by the Society: ASTM Practice for Testing Biomaterials in Rabbits for Primary Skin Irritation (F 719) and ASTM Practice for Testing Guinea Pigs for Contact Allergens: Guinea Pig Maximization Test (F 720). An additional five documents appeared for the first time in the 1982 edition of the *ASTM Book of Standards*. These included an ASTM Practice for Evaluating Extracts by Intracutaneous Injection in the Rabbit (F 749), Practice for Evaluating Material Extracts by Systemic Injection in the Mouse (F 750), Practice for Assessment of the Hemolytic Properties of Materials (F 756), and Practice for Short-Term Screening of Implant Materials (F 763). The fifth document was ASTM Practice for Selection of Generic Biologic Test Methods for Materials and Devices (F 748).

This document (F 748) addressed the issue of what types of tests should be used to demonstrate sufficient safety of materials for particular end use applications. The matrix approach is used similar to that used by Health Industries Manufacturers Association/Pharmaceutical Manufacturers Association (HIMA/PMA) and the British Standards Institution in which end use applications were broken down according to length of time, and types of tissue contact were listed in rows, and particular test methods were listed in columns permitting indication of what types of tests were considered sufficient for particular levels of end use application. A common theme in all of the applications was that the first stage of biocompatibility screening would be some sort of acute in vitro cell-culture toxicity test for identifying materials with toxic extractable moieties.

The subject of cell-culture toxicity has, as with the other methods of biological testing, been one of active debate over the past six to eight years. While there are many industries and commercial laboratories that are currently using cell-culture toxicity methods for screening and testing materials, there is no real consensus as to which methods should be used. There are several ways of conducting a cell-culture test (for example, direct contact, agar overlay, and so forth), and there are a number of different cell types that may be used in any of these methods. During the years of discussing standardization of cell-culture testing, it became apparent that different laboratories were using different methods for good reasons. The nature of the technique of cell-culture testing per se and the nature of the product being tested resulted in the selection of a particular method and a particular cell line. Many task force meetings seemed to repeat themselves on the subject of "who was doing what and why." At the Nov. 1981 meeting of ASTM Committee F-4, it was decided that the way to resolve the conflicts was to organize a symposium at which laboratories could present their methods and rationales for use of methods, as they pertain to their particular product. The scope of the meeting was laid forth in the letter sent to speakers as an invitation to participate.

> The biocompatibility section of ASTM Committee F-4 on Medical and Surgical Materials and Devices has been struggling for several years with the task of writing standardized methods for cell culture testing of implant materials. It has become clear that there are a number of labs that are using different methods, different cell lines and are doing the testing for different purposes. The approach we would like to take is to develop standards for several methods and to develop recommendations as to when or for what reason a particular test should be used. In order to sort out the differences and to develop standards and recommendations, we are organizing a one-day meeting to which the major labs performing these tests are to be invited to present their methods and rationales for use.
>
> We would like you or one of your technical people with knowledge of the details of cell culture testing to make a presentation and to participate in a workshop on this subject. The presentation should discuss the details of what you are doing (type of test, cell line, etc.) and for what level of testing (initial screen, lot-to-lot, etc.). We would like to know what works for you and why. As part of this we ask that you or your representative also write a paper addressing these issues, which we can use for the workshop, subsequent standards development, and which will also be included in an ASTM publication of the meeting and workshop.

The one-day symposium was organized with presentations being given in the morning and the first part of the afternoon. The second part of the afternoon was devoted to a 2-h open discussion workshop to review and discuss the papers that we had presented, as well as to discuss the role of ASTM Committee F-4 on the development of cell-culture standard testing methods.

The papers presented in this volume are by individuals from laboratories using cell-culture test methods for biomaterial testing and screening. The papers represent, in many instances, commercial input and commercial interest. Their intent is not to say which method is correct, but rather to state what method works for the particular company with its particular products. The papers are organized with the first series discussing specific types of methods that are being used and their particular end use applications, followed by several papers that addressed the most critical question of the correlation between in vitro cell toxicity test results and in vivo performance of materials. Two papers address specific end use application and discuss using radioactive tracers that permit more quantitative assessment of the cellular response to materials or extracts. The final paper discusses the availability and standardization of the cells that are actually used for cell-culture testing. Finally, the proceedings from the workshop are presented in the back of this volume. The text and the workshop has received minor editing for clarity, but essentially represents a free discussion of the issues being addressed. The workshop is organized into four parts, the first three addressing specific questions put to the audience by the Chair: (1) What should Committee F-4 be doing about cell-culture test standardization? (2) Should we develop standard charts or methods for ranking cell reactions? (3) How do we address the problem of standard reference materials for cell-culture testing? (4) And a question and answer period where questions that had been written to specific speakers during the morning session were answered and discussed.

The consensus that came from the workshop discussions was that the role of Committee F-4 was to develop documents that could be referred to as "benchmark" standards. While these might not be the specific methods being used on a day-to-day basis by all laboratories, they would be methods that could be used for validating and comparing different methods. The adopting of methods by Committee F-4 was not to be interpreted as an endorsement of anyone's particular method, but rather the development of specific methods that could be standardized and used.

The mood of the biocompatibility section and the subsequent task force meetings was that with this frame of mind we could move forward in developing standardized test methods. Perhaps as a direct result of this symposium the biocompatibility section completed development of its first cell-culture method, which was passed to society ballot at the Nov. 1982 meeting of Committee F-4. It is anticipated that the committee will develop several methods for cell-culture cytotoxicity testing and screening materials over the next few years. With standardized methods, we will be able to move forward in developing a better understanding of the applicability of particular methods, and of the accuracy and reliability of in vitro methods predicting in vivo performance.

As editor and chairman, I would like to thank Thomas Nickel of Travenol Laboratories and Thomas Walker of the Canadian Standards Association for their assistance in organizing the symposium.

Stanley A. Brown
University of California, Davis, Calif. 95616, symposium chairman and editor.

Linda M. Smith[1]

Direct-Contact Cell-Culture Method

REFERENCE: Smith, L. M., **"Direct-Contact Cell-Culture Method,"** *Cell-Culture Test Methods, ASTM STP 810*, S. A. Brown, Ed., American Society for Testing and Materials, 1983, pp. 5–11.

ABSTRACT: A direct-contact cell-culture test method has been developed to provide routine lot-to-lot biological assurance as a part of the quality assurance program during the manufacture and distribution of medical materials and sterile-packaged human-implant devices. It is particularly useful for detecting trace ethylene oxide or catalyst residuals. The method has also been used routinely to provide a first biocompatibility evaluation of proposed new materials or formulation.

Specimens are placed in direct contact with human fibroblastic cells grown to a confluent monolayer. The results are evaluated mircroscopically after 24 h incubation, using appropriate positive and negative controls. An alteration in cellular morphology is considered a cytopathic effect (CPE). Implant materials and devices are approved by the quality assurance department only if they produce no CPE in this test.

This test method has been used continuously in the lot-to-lot quality assurance program at Dow Corning's Medical Products Plant for more than ten years and has provided a fast, reliable, highly sensitive, and inexpensive method for evaluation of biocompatibility.

KEY WORDS: tissue culture, direct contact, biocompatibility, cytopathic effect, medical devices, cell culture, cytotoxicity

Cell culture can be considered an extremely practical and sensitive method of investigating acute biocompatibility for the prediction of toxicity potential [*1*].[2] In our laboratory at Dow Corning in the early 1970s, the use of a direct-contact cell-culture method was initiated for the evaluation of implantable medical devices on a lot-to-lot basis. Over the more than ten years that this method has been used routinely, we have developed a great deal of confidence in its reliability and reproducibility. The use of cultured human diploid cells in a rapid, cost-efficient in vitro test system has provided an effective alternative to animal implant testing. This paper will outline the detailed method of how the test is performed currently in our laboratory, reflect on some of the developmental work that has determined why it is done the way it is done, and, finally, discuss the usefulness of this test in our daily operations.

[1] Quality assurance biology laboratory supervisor, Dow Corning Corp., Midland, MI 48640.
[2] The italic numbers in brackets refer to the list of references appended to this paper.

Method

The direct-contact test-method [*2,3*] currently uses WI-38 human embryonic lung cells (Flow Laboratories, McLean, Va.) and HR-218 human foreskin cells (HEM Research, Rockville, Md.). Two cell lines are used routinely so that a backup is always available in case of problems with one line. Other human cells have been used after qualification to ensure that their sensitivity is equivalent to the previously mentioned lines. All cells are received and propagated in 150-cm^2 flasks that are maintained in a 5% carbon dioxide humidified atmosphere at a temperature of 37 ± 2°C. Cells for biocompatibility testing are seeded from the larger flasks into 35 × 10-mm cell-culture wells and usually reach monolayer within three to four days. Seeding is done by a routine trypsinization procedure. This entails rinsing of the cell sheet with Hank's balanced salt solution (modified) without magnesium and calcium (Flow Laboratories, Rockville, Md.), cell dispersion using 0.5% trypsin at 37°C, centrifugation to harvest the cells, centrifugation to wash the cells, and, finally, seeding by the addition of five to seven drops of cell suspension to the cell-culture wells containing fresh medium. Planting of additional 150-cm^2 flasks for inventory can be also done at this point.

When the cells in the test wells are grown to a confluent monolayer, as confirmed microscopically, the medium is aspirated and replaced with a minimal amount (1.5 to 2.0 mL) of fresh medium. Minimum essential medium with Earle's salts, 10% fetal bovine serum, and 5 mg/500 mL L-glutamine (Gibco Laboratories, Grand Island, N.Y.) is used routinely. Earle's basal medium (BME) can be also used. Specimens to be tested are gently but firmly placed directly on the monolayer of cells, and the wells are incubated 24 h at 37°C and 5% carbon dioxide. Each group of specimens must include a positive and a negative control. A good, reliable positive control is a freshly cut surface of clean, sterilized, latex rubber. The negative control is a Dow Corning® brand MDX4-4515 silicone elastomer cured 6 min at 155°C and post cured 2 h at 250°C. This negative control is a product that has a long history of biocompatibility based on a large amount of historical data.

After 24 ± 1 h incubation, each cell well is examined at ×10 magnification for any change in cellular morphology. Any alteration in morphology, ranging from focal clumping and granulation of the cells to complete destruction of the cell sheet, is considered a cytopathic effect (CPE). Figures 1 and 2 show WI-38 cells with no cellular change (Fig. 1) and with a cytopathic effect (Fig. 2), after 24 h of contact with a specimen. Material being tested is reported as producing either "no CPE" (no change in the cell morphology) or "CPE" (any alteration in cell morphology). An alternative grading system from one to five may be also used, in which one is the least amount of cellular change, and five is the greatest amount. A specimen that has produced a CPE can be retested using two specimens of the material. If neither of the second series produces a reaction, the test is reported as "no CPE" (retest was required).

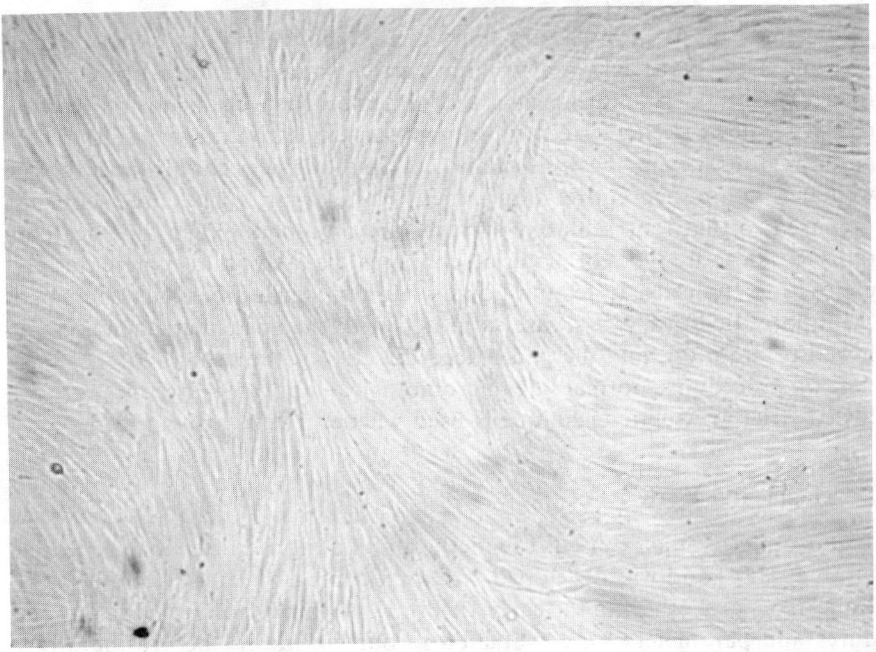

FIG. 1—*WI-38 human embryonic lung cells (×10 magnification) showing no cytopathic effect.*

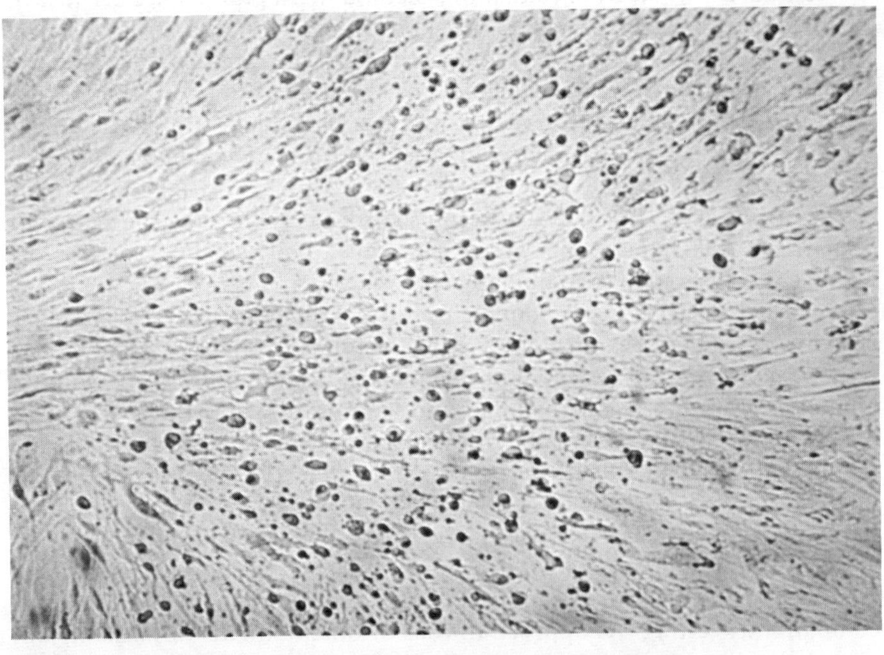

FIG. 2—*WI-38 human embryonic lung cells (×10 magnification) showing a cytopathic effect.*

Specimen preparation is very important. Specimens must be as clean as possible, free of any bacterial contamination or particulate debris that can interfere with the test. Specimen size is not critical as long as it is representative of the device being tested and enough space is left in the well so that cells adjoining the specimen can be examined. Small devices that fit into the cell well are used in their entirety. Specimens from larger devices are aseptically prepared so that representative surfaces are obtained. When a device contains two or more types of material, care must be taken that each material is exposed to the cells. In testing liquids, 10-mm sterile fiberglass filter pads (Millipore, Bedford, Mass.) are saturated with the specimen, the excess allowed to drain off, and the pad placed directly on the cell.

In all procedures, normal safe microbiological laboratory practices are followed strictly. Aseptic technique is used whenever cells are handled.

Developmental Work

Our direct-contact cell-culture method was designed after that of Rosenbluth et al [*4*] and was developed at Dow Corning by E. A. Abbott, G. W. Robertson, and others in the early 1970s [*5*]. A number of varied test procedures and parameters were evaluated in our laboratory before the direct-contact technique was deemed to be the method of choice. Chief among the other procedures evaluated were many variations of the agar/agarose overlay method. It was determined that this method was a poor choice because of the inability of the agar or agarose to allow diffusion of organosilicones and other nonaqueous or fat-soluble materials [*6*]. This greatly decreased the sensitivity because the test is dependent on the ability of materials to diffuse through water-soluble agar to the cell surface. By comparison, in the direct-contact method all the specimen components are directly available to the cell layer.

The principle cell line that has been used routinely through the years has been the WI-38 human embryonic lung cells. This cell line is the most highly characterized normal human-cell population known and is the standard by which all other human diploid cell strains are compared. Concern has been expressed in recent years concerning the future availability of the WI-38 cell line; however, this potential problem has not yet manifested. Any other human-cell line that is used in our laboratory must first undergo a rigid qualification procedure to ensure that it is as sensitive and as reproducible as the WI-38. Other cell lines that have been used include HR-218 human foreskin, MRC-5 human embryonic lung, and IMR-90 human embryonic lung. All these cell lines have been shown to have virtually identical sensitivity as the WI-38 cells through duplicate testing of over 400 production specimens. Our laboratory has always used diploid cell populations of human origin for biocompatibility testing. As opposed to cells of other animals or of malignant origin, these human cells contain a normal diploid chromosome count, do not contain a foreign genome, mycoplasma, or virus, and, most importantly, originate from the human species. In order to design this in vitro procedure as closely to the in vivo situation as possible, the use of human diploid cell lines appears warranted.

Over the years, our laboratory and other laboratories have confirmed the sensitivity of tissue-culture techniques versus animal testing [*7*]. Over 500 specimens of various materials have been evaluated at Dow Corning concurrently in animal systems and in cell culture. Without exception, we have found that implant-reactive materials always elicit a cytopathic response in cell culture. However, because of the degree of sensitivity, reactive materials in cell culture are not always apparently reactive in the in vivo tests. These two facts make the cell-culture technique a valuable screening tool to indicate what might be expected in higher testing.

It has been also shown by work done at Dow Corning that this cell-culture method can be used to detect known levels of ethylene oxide (ETO) residuals in gas sterilized devices. During this developmental work, the direct-contact cell-culture technique was used in conjunction with gas chromatography to monitor the dissipation of the residual [*5*]. Biocompatibility on WI-38 human embryonic lung cells was correlated with the amount of residuals in the sterilized specimens. A great deal of investigating was done concerning clearance rates of residual chemical as compared to the nature of the material, size of the device, and the environmental conditions of aeration. Tables 1 and 2 show typical dissipation data as compared to CPE in cell culture. Although the sensitivity of the cell-culture test will vary with the product that is being evaluated, particularly with the rate at which the product releases the ETO, we can reliably estimate the detection limit of ETO in our medical products by direct-contact cell-culture to be approximately 50 ppm. Diluted ETO gas placed directly into the growth media over cells has elicited a lower response range of 36 to 45 ppm. This work has enabled us to design dissipation systems for each type of device depending on the specific clearance rates determined by this work and to monitor the efficacy of these dissipation systems by routine cell culture.

TABLE 1—*ETO dissipation versus tissue-culture reaction (cleared at 53°C [127°F]).*

Time After Sterilization, h	ETO Concentration, ppm	Tissue-Cell Culture Reaction
0	920	CPE
6	670	CPE
24	460	CPE
50	180	CPE
72	120	CPE
98	44	no CPE
116	24	no CPE
168	5	no CPE

TABLE 2—*ETO dissipation versus tissue-culture reaction (cleared at 56°C (132°F)/partial vacuum).*

Time After Sterilization, h	ETO Concentration, ppm	Tissue Cell-Culture Reaction
0	4 100	CPE
2	2 700	CPE
7	2 300	CPE
24	214	CPE
30	62	CPE
48	5	no CPE
72	ND[a]	no CPE

[a] ND = not detectable.

Uses

The direct-contact cell-culture test is used extensively at Dow Corning, both in our day-to-day production operations and in new product development. All of our sterile-packaged medical devices must pass the direct-contact cell-culture test without a CPE on a lot-to-lot acceptance basis. This enables us to monitor both the biocompatibility of the finished device as well as the ETO residuals. A production lot of sterilized products that fails the cell-culture test must be reevaluated for ETO content, sterility, or biocompatibility of both the finished products and the basic materials.

Lot-to-lot cell culture is performed routinely on all of our bulk elastomers and gels. This enables us not only to be able to certify biocompatibility to our customers who buy these medical materials but also to maintain control of our production processes. In the same light, many of our raw materials and intermediates are also tested routinely so that a potential problem may be discovered and corrected as early as possible.

Cell culture is also used to monitor the process development of a new product or production method. We use cell culture as a method enabling us to follow chemical reactions of systems that contain components that in themselves may produce a CPE if unreacted in the final formulation. Interval specimens are taken to determine when the cytologically active component of the formulation is sufficiently reacted chemically to not elicit a response in cell culture. This can tell us not only the biocompatibility of a final product but also when the chemical reactions involved are at or nearing completion. Reliability engineers can use this sensitive technique as a part of their "trouble-shooting" tests to determine the cause of both developmental and production problems.

Conclusions

The versatility of this relatively simple and inexpensive cell-culture method makes it an important tool for us to monitor the biocompatibility of both our production processes and our final products. We believe that it the most sensitive of the qualitative cell-culture tests available today and, as such, should be used as a standard for those devices requiring the most stringent of testing, that is, long-term implant materials.

Acknowledgments

I wish to thank E. A. Abbott and L. M. Veresh for their cooperation and their generosity with both their time and their data, both past and present, for use in this paper.

References

[*1*] Adamas, R. L. P., *Cell Culture for Biochemists*, Elsevier/North-Holland Biomedical Press, Amsterdam, The Netherlands, 1980.

[*2*] Corporate Test Method 0274, "Tissue Cell Culture—Fluid Medium—Direct Contact", Dow Corning Corporation, 1972, Midland, Mich.

[*3*] Gill, M., "Direct Cell Contact Screening for Materials and Devices," *Medical Device and Diagnostic Industries*, Canon Communications, Inc., Los Angeles, March 1982, pp. 72–76.

[*4*] Rosenbluth, S. A., Weddington, G. R., Guess, W. L., and Autian, J., *Journal of Pharmacological Science*, Vol. 54, 1965, pp. 156–159.

[*5*] Abbott, E. A., Dow Corning Corporation, Midland, Mich., personal communication.

[*6*] Lord, G. H., "Bio-evaluation of Materials and Devices: Philosophical, Rational and Specific Considerations," presented at The Biomaterials Session of the Gordon Research Conference, Proctor Academy, N.H., 1976.

[*7*] Wilsnack, R. E., Meyer, F. J., and Smith, J. G., "Human Cell Culture Toxicity Testing of Medical Devices and Correlation to Animal Tests," Biomat., Med. Dev., Art. Org., Vol. 1, No. 3, March 1973, pp. 543–562.

Linda M. Ettorre[1] *and Kathy Phillips*[1]

Test Method for Human Cell-Culture Elution to Evaluate Medical Materials and Devices for Cytotoxicity

REFERENCE: Ettorre, L. M. and Phillips, K., **"Test Method for Human Cell-Culture Elution to Evaluate Medical Materials and Devices for Cytotoxicity,"** *Cell-Culture Test Methods, ASTM STP 810,* S. A. Brown, Ed., American Society for Testing and Materials, 1983, pp. 12–18.

ABSTRACT: Tissue-culture testing used for screening selected new-components as well as for lot-to-lot testing of designated incoming materials before approval for use is performed routinely by the employees of the Microbiology Department at Extracorporeal, Inc. The "extraction method," which uses the WI-38 cell line, is used most often. MRC-5 cells, a similar cell line, are also maintained as an alternate.

Materials are extracted on a surface area or weight ratio basis at 37°C for 24 h. The growth media is aspirated from the plates, replaced with extracting media, and incubated at 37°C for 24 h. Plates are stained, dried, and examined under a light microscope. Cytotoxicity is scored on a scale of 0 (negative response) to 4 (more than 75% of cells morphologically changed). Results of +2 to +4 are considered cytotoxic. Retesting is done to verify cytotoxic results.

KEY WORDS: cell culture, extraction, WI-38, cytotoxicity

Cell-culture test methods have been used for over ten years to evaluate the cytotoxicity potential of medical materials and devices. Today several different cell lines are used employing different techniques. Standardization of these test methods is necessary in order to obtain reproducible results among laboratories.

1.0 Scope

1.1 This proposed standard describes a method of human cell-culture elution to evaluate the cytotoxic potential of medical materials and devices.

1.2 This method is one of a series of recommended practices having the same purpose but using different cell-culture techniques.

[1] Research associate and microbiologist, respectively, Extracorporeal, Inc., Ross and Royal Roads, King of Prussia, PA 19406.

2.0 Applicable Documents

2.1 Wilsnack R. E., Meyer F. J., Smith J. G., "Human Cell Culture Toxicity Testing of Medical Devices and Correlation to Animal Tests," *Biomaterials, Medical Devices, and Artificial Organs,* Vol. 1, No. 3, 1973, pp. 543–562.

2.2 The American Type Culture Collection (ATCC), *Catalogue of Strains II,* Third ed., H. D. Hatt, Ed., ATCC, Rockville, Md., 1981.

2.3 United States Pharmacopeial Convention, Inc. (USPC), *The United States Pharmacopeia,* 20th Revision and *The National Formulary,* 15th ed., USPC, Rockville, Md., 1980.

3.0 Summary of Method

Cell cultures are grown to a confluent monolayer in tissue-culture dishes. The growth media is aspirated and replenished to provide a resting confluent cell layer. The test material/device and negative and positive controls are extracted with media. Growth media is aspirated from tissue-culture dishes and replaced with test and control material/device extracts. Tissue-culture dishes are incubated, stained, and examined under a microscope for cytotoxicity.

4.0 Significance

4.1 This cell-culture test method is useful for screening new material formulations for cytotoxicity as well as a lot-to-lot control program for medical materials and devices.

4.2 This test method assumes that assessment of cytotoxicity predicts the potential for adverse biological reactions to medical materials and devices during clinical applications to humans.

4.3 This cell-culture test method is suitable for use in specifications and standards for medical device materials and devices.

4.4 Some biomaterials with a history of safe clinical use in medical devices are cytotoxic. This recommended practice does not imply that all biomaterials must pass this assay to be considered safe for clinical use.

5.0 Apparatus

The following apparatus will be used:

(*a*) Incubator to maintain a temperature of 37 ± 1°C, an atmosphere of 5 ± 1% carbon dioxide, and greater than 90% humidity.

(*b*) Water bath to maintain a temperature of 37 ± 2°C.

(*c*) Laminar flow work area capable of filtering out particles greater than 0.3 μm in diameter.

(*d*) Borosilicate bottle (150 mL), United States Pharmacopia (USP) Type I (Wheaton "400" Brand or equivalent).

(*e*) Sterile, disposable centrifuge tube (Corning 25330-50 or equivalent).

(*f*) Inverted optical microscope, with magnification of ×40, ×100, and ×200.

(*g*) Hemacytometer.

(*h*) Sterile, disposable 35 by 10-mm tissue-culture dishes (Corning 25000-35 or equivalent).

(*i*) Sterile, disposable 150-cm^2 polystyrene tissue-culture flasks (Corning 25120-150 or equivalent).

(*j*) Centrifuge capable of attaining 1000 g.

(*k*) Sterile, disposable pipettes (1, 5, and 10 mL).

6.0 Reagents

6.1 The following reagents will be used:

6.1.1 Low suds automatic dishwasher cleaning solution.

6.1.2 Tissue-culture glassware cleaning solution.

6.1.3 5% sodium hypochlorite solution (Clorox® bleach).

6.1.4 50-mL minimum essential medium (MEM, Eagle, ×10 concentrate) with Earle's salts, without L-glutamine and $NaHCO_3$ with:

(1) 60-mL newborn calf serum.

(2) 6-mL penicillin-streptomycin-Fungizone®.

(3) 6-mL nonessential amino acids (×100).

(4) 6-mL L-glutamine (×100, 200 m*M*).

(5) 0.3-mL Garamycin/500 mL MEM.

(6) 13.5-mL sodium bicarbonate.

(7) 435-mL water.

6.1.5 Trypsin (0.25%, ×1).

6.1.6 Dulbecco's phosphate buffered saline (D-PBS, ×1).

6.1.7 Hematoxylin stain solution with 30-mL isopropyl alcohol added per 0.45-kg (16-oz) stain.

6.1.8 Methanol.

6.2 All reagents shall be tissue-culture grade.

6.3 Reagents shall be reconstituted in accordance with the manufacturer's directions, using aseptic technique.

6.4 Reagents shall be stored in accordance with the manufacturer's directions unless otherwise indicated in 6.1.

7.0 Cell Cultures

Cell cultures used in this assay shall be of the WI-38 Cell Line CCL75-American Cell Type Collection.

NOTE: Other cell lines may be used, if data for five materials known to be cytotoxic and five materials known to have no cytotoxic effect are available to relate the sensitivity of the chosen cell line with that of the WI-38 cells.

8.0 Control Materials

8.1 A negative control shall be prepared in accordance with Section 12.0, using materials previously known to yield a negative response. Medical grade silicone elastomer is generally used.

8.2 A positive control shall be prepared in accordance with Section 12.0, using a material previously known to yield a +4 response as defined in Section 13.1. Latex tubing is generally used.

9.0 General Technique

9.1 Aseptic technique is used for all reagent preparation and culture or medium transfer.

9.2 All solutions are warmed to a temperature of 37 ± 1°C before use.

9.3 All glass vessels used during the assay are washed with a cleaning solution and rinsed thoroughly with deionized water with maximum conductivity of 80 μS.

9.4 All reagents, culture vessels, pipettes, and other materials must be sterile before use.

9.5 All cell manipulations are performed under a laminar flow hood.

9.6 Laminar flow hood and incubator trays are cleaned with 5% sodium hypochlorite solution before use.

10.0 Specimen Preparation

10.1 Specimens should be as consistent with the end-use product as possible, including washing procedures, sterilization, and so forth.

NOTE: Materials subjected to ethylene oxide (EO) sterilization should be allowed to degas for at least 48 h so that EO residuals do not contaminate the test.

10.2 There are no specified dimensions for each specimen.

10.3 If a specimen is constructed of more than one material, the whole specimen is tested. If this is not possible, separate specimens of each material are tested.

11.0 Cell-Layer Preparation

Confluent cell monolayers will be prepared as follows:

11.1 Aspirate the medium from a 150-cm^2 cell-culture flask containing a confluent cell monolayer.

11.2 Rinse the cell layer with approximately 10 mL D-PBS. Aspirate the D-PBS rinse.

11.3 Wash the cell layer with 2 ± 0.1 mL trypsin solution (0.25%). Aspirate the trypsin rinse.

11.4 Add 1.5 ± 0.1 mL trypsin (0.25%). Thoroughly wash the cell layer.

11.5 Incubate for 5 to 10 min to suspend the cells.

11.6 Add 8 ± 0.5 mL culture medium to the flask. Mix the suspension thoroughly to completely separate cells.

11.7 The suspension is divided equally between two new 150-cm^2 flasks. Approximately 40-mL MEM is added to each flask. Each flask is incubated until a confluent monolayer of cells has formed.

11.8 To prepare 35-mm test plates, dilute the cell suspension (in Section 11.6) to 6×10^4 cells/mL. A hemacytometer can be used for counting the cells. Using a sterile pipette transfer 2 ± 0.1 mL of this cell suspension to each test plate.

11.9 Incubate until a confluent monolayer is formed, usually five to six days.

12.0 Test Procedure

12.1 Material specimens are tested using a 1:5 weight (grams) to volume (millilitres) ratio. Materials for which the surface area is easily calculated are tested according to procedures of the United States Pharmacopeial Convention, Inc. (USP XX, p. 951, Table 4).

NOTE: End use of the material and its relationship to patient exposure should be considered.

12.2 Place each specimen in a sterile borosilicate container with the appropriate amount of MEM.

12.3 Extract all material specimens for 24 ± 1 h at 37 ± 1°C.

NOTE: Methods of extraction for device specimens are developed in accordance with the product end use. No standard guidelines are available.

12.4 If, at the end of the extraction period, there is particulate matter present, the MEM extract is decanted and centrifuged for 10 min at 400 to 1000 g.

12.5 Microscopically examine test plates for confluency.

12.6 Aspirate growth medium from test plates and replace with 2 mL of test extract. Prepare duplicate plates for each specimen.

12.7 Incubate the plates for 24 ± 1 h at 37 ± 1°C with 5 ± 1% carbon dioxide with humidity greater than 90%.

12.8 Stain plates as follows:

(*a*) Discard extraction medium.
(*b*) Rinse each plate with 2-mL D-PBS and discard.

(*c*) Add 2-mL methanol to each plate. Let stand 20 min.

(*d*) Discard methanol.

(*e*) Place 2-mL hematoxylin stain on each plate. Let stand 40 min.

(*f*) Rinse each plate in warm tap water. Invert on paper towel and air dry thoroughly before microscopic examination.

13.0 Evaluation of Results

13.1 All plates are examined microscopically and scored with an inverted microscope at magnifications of ×100 and ×200. Cytotoxicity scores range from 0 to +4:

0 = negative response, excellent outgrowth, normal cell morphology; little or no cellular detachment or debris.

+1 = outgrowth comparable to control, but increased intracellular granulation, *or* less than 25% reduction of cells or morphological change (slight swelling or crenation) of individual cells without presence of granulation.

+2 = distinct cellular swelling or crenation affecting nearly all the cells in the monolayer, usually accompanied by intracellular granulation; floating debris may be present.

+3 = extensive granulation *or* swelling or crenation, or both, resulting in the loss of 50 to 75% of cells.

+4 = loss of more than 75% of cells *or* severe morphological change; the culture may still exhibit metabolic activity.

13.2 A cell culture shall be deemed to show a cytotoxic effect if microscopic examination reveals a score of +2, +3, or +4. A retest of the test material shall be performed to confirm cytotoxicity.

13.3 If, for a given set of specimens, a cytotoxic effect is observed for the negative controls, or no cytotoxic effect is elicited by the positive controls, the results for that test shall be considered invalid.

14.0 Report

The report of the test results shall include at least the following:

(*a*) test date,

(*b*) extraction date,

(*c*) cell strain used,

(*d*) medium used,

(*e*) a description of all test and control specimens including:

 (1) material or device name,

 (2) lot or batch number,

 (3) material or device manufacturer,

 (4) size and method of preparation, and

(5) method of sterilization;

(*f*) a statement whether each test material is nontoxic or cytotoxic; and

(*g*) any other pertinent observations including any deviations from the method described herein.

15.0 Precision and Accuracy

Quantitative precision and accuracy have not been determined.

Summary

In summary, human cell-culture elution using WI-38 cells, is useful for predicting the cytotoxic potential of medical materials and devices. Standardization of this and other test methods is necessary in order to obtain reproducible results among laboratories and for their inclusion in medical device specifications and standards.

Judy Fehn[1] *and Denise Schottler*[1]

Tissue-Culture Methods for Determining Biocompatibility

REFERENCE: Fehn, J. and Schottler, D., "**Tissue-Culture Methods for Determining Biocompatibility,**" *Cell-Culture Test Methods, ASTM STP 810,* S. A. Brown, Ed., American Society for Testing and Materials, 1983, pp. 19–24.

ABSTRACT: Tissue-culture test methods are used as acute screening for biocompatibility testing of materials and solutions used in the manufacture of our biomedical devices at Medtronic, Inc. Presently, we are capable of performing three methods, they are: agar overlay, minimal essential medium (MEM) elution, and low cell density. In all the test methods performed we use L-929 mouse fibroblasts as our cell line.

Tissue-culture procedures are used for screening the safety of materials that are being evaluated in the following situation: (1) new material and vendor qualification, (2) process change or qualification or both, (3) vendor audit, and (4) incoming inspection. Medtronic basically follows the guidelines of the Health Industry Manufacturers Association (HIMA) Report 78-7, which classifies devices into four categories based upon usage. These are intended to be guidelines, but ultimately the criteria to classify materials and the testing done on the material are dependent upon (1) the intended end use of the material and its application, (2) whether it has direct or indirect body tissue, or fluid contact or both, (3) whether it will be short term or long term or both, and (4) if the material is put through a process that may change or affect the material or both.

The decision as to which tissue-culture method to use by which to screen a material is determined by (1) the nature of the material, (2) information required of testing, and (3) history of material and which test method is most applicable. The results of tissue-culture testing are meant to be screening and do not replace in vivo testing. If the results of a tissue-culture test are questionable, it is the decision of the study director of how to proceed. Medtronic has been doing tissue-culture testing for more than ten years and has a good data base to support our test methods.

KEY WORDS: agar overlay, minimal essential medium elution, low cell density, screening methods, end use, application, cell culture

Medtronic is the world's leading producer of implantable medical devices. The company designs, manufactures, and markets heart pacemaking systems, neurological devices, mechanical heart valves, and computer systems for diagnostic imaging. Medtronic also provides instruments and services for cardiac monitoring. This diverse range of products demands a diverse program for biocompatibility screening.

[1] Supervisor in laboratory service and in vitro toxicology laboratory technician, respectively, Physiological Research Laboratories, Division of Medtronic, Inc., Minneapolis, Minn. 55433.

Medtronic has been doing essentially all its own biocompatibility testing for more than ten years. During this time, we have developed a battery of in vitro and in vivo test procedures. The in vitro test procedures include the tissue-culture test methods, which are used as acute screening for biocompatibility testing of materials and solutions used in the manufacture of our biomedical devices.

Procedure

For rapid screening of biomaterials, cell death and inhibition of cell growth are the simplest and most frequently used parameters. Presently we use three tissue-culture methods for testing those parameters using L-929 mouse fibroblasts as our cell line.

The tissue-culture methods are:

1. *Agar Overlay*—This is our primary test method for screening materials and solutions. We essentially follow the methodology developed by Guess [*1,2*].[2] The method involves growing a monolayer of mouse fibroblasts, removing the liquid phase media, and replacing it with a solid agar phase media. Two specimens, a negative and a positive control, are placed on each plate and incubated 24 h at 37°C in a 95% air, 5% carbon dioxide atmosphere. After 24 h, the cells are stained with a vital stain (neutral red) and examined microscopically for cell lysis and the extent of the lysis zone. These two criteria are rated on a scale of 0 to 5 with 0 being no cytotoxic response, and 5 being the highest cytotoxic response (or the most cytotoxic). The toxicity of the specimen is reported as a response index, which is determined by the following formula:

$$\text{response index} = \text{lysis zone index}/\text{cell lysis index}$$

The lysis zone index is the area under and around the specimen affected by the material. The cell lysis index is the percentage of cells lysed within the lysis zone.

2. *Extraction/Minimum Essentials Medium* (*MEM*) *Elution Method*—The methodology was developed from the works of Wilsnack [*3,4*] and is used primarily when the material cannot be tested by the agar overlay method or when a more sensitive test is called for or for both reasons.

The MEM elution method again calls for a confluent layer of mouse fibroblasts plated in the 6-mm wells of a microtest culture plate. The specimen being tested is eluted in an extraction tube containing a precise amount of MEM. Appropriate positive and negative controls as well as a media negative are eluted at the same time. Specimens are eluted for 24 h at 37°C. The

[2] The italic numbers in brackets refer to the list of references appended to this paper.

cytotoxicity of the specimen is determined by preparing twofold dilutions (1:1 to 1:128) of the specimen eluates in MEM. An aliquot of each dilution is applied to the prepared cells from which the media has been removed.

Plates containing sample eluates (each dilution run in triplicate) are again incubated for 24 h at 37°C and 95% air/5% carbon dioxide atmosphere. After incubation, the plates are decanted and the media replaced with Trypan blue, an exclusion dye that stains only damaged, nonviable cells. The plates are then examined microscopically, and the cells counted. A cytotoxic score, determined from the following formula, is assigned to each well counted.

$$\text{number of lysed cells/total number of cells} = \text{\% of lysed cells}$$

A material extract or dilution of the extract is considered toxic when the percentage of cell growth in the extract (or dilution) is $<50\%$ or cell growth in the negative control. A $\geq 50\%$ cell growth is considered passing. If the objective of the study is to determine the dose response to chemical dilutions, the concentration that produces 50% cell growth is reported.

3. *Low Cell Density*—The methodology was developed from the works of Wilsnack [*5*]. This method is used for two applications:

a) It evaluates the materials biocompatibility by comparing the number of fibroblast colonies formed in control medium to those colonies formed in material extract.

b) This method also evaluates the effect a chemical may have on a mouse fibroblast colony formation. This is done by exposing the cells to a series of dilutions of a chemical to determine at what level colony formation is inhibited by 50%, thus giving dose response data.

This test is the most sensitive of the three currently in use, and it is used almost exclusively when there is a need to test cell response to exact dosages of a test material, usually over a wide range of dosages. The results are then plotted percent specimen against percent cell growth to obtain a toxicity curve for the specimen.

This test procedure involves growing the L-929 fibroblast cells in 60- by 15-mm dishes with millimeter grids. The cells are very lightly plated and are not grown to confluency before use as one of the prime objectives of this test is to study inhibition of cell growth over a longer period of time (eight to ten days).

Each different dose of specimen extract is applied in triplicate. After 72 h of incubation at 37°C and 95% air, 5% carbon dioxide atmosphere, the first dose is removed and replaced by a second dose of the same extract. This remains on the cells for an additional five to seven days until cells in the negative control plate have grown sufficiently to evaluate colony growth. The plates are then decanted and stained with Trypan blue, which is only taken up by damaged or nonviable cells.

The response to test material dilution is evaluated by counting five of the 1-mm square grids in each of the three plates, averaging the total of all grids counted (for each dilution). This provides the colony average per grid, which is then divided by the colony average of the negative control to obtain the percentage of cell growth for each dilution. The percentage of cell growth is plotted against the percentage of test material per dilution, which provides us with a toxicity curve for the specimen. The objective of the study is to obtain the range over which a concentration of test material will produce 50% cell growth.

Discussion

Medtronic basically follows the guidelines of the Health Industry Manufacturers Association (HIMA) Report 78-7 [*6*], which classifies devices into four categories based upon usage. These categories are meant to be guidelines but ultimately the criteria used to classify test materials and testing required of these materials are as follows:

(1) the intended end use of the material and its application,
(2) whether it has direct or indirect body tissue or fluid contact or both,
(3) whether it will have short- or long-term usage, and
(4) if the material, while during manufacturing, is put through a process that may alter or affect the material or both.

Because of the variety of our biomedical products all the materials used in their manufacture are evaluated by this criteria. Therefore in some cases a material may have multiple uses and may have to meet a different set of criteria each time.

The decision as to which tissue-culture method to use to screen a material or solution is determined by

(1) the nature of the material being tested;
(2) information required from the test procedure, whether we are looking for degradation, dose response, toxicity curve, and so forth;
(3) the history of the material being tested, and which test method is the most applicable.

The results of tissue-culture testing are primarily meant to be a screening procedure that can be at least partially correlated to existing data derived from both in vitro and in vivo animal testing. Unless segregated as to parameters and type of test material, comparisons between animal tests and tissue-culture tests are invalid as are comparisons between different cell-culture tests. Although cell-culture tests more frequently detect toxicity than do animal tests, the results of the tissue-culture tests should not be considered as falsely positive because this very sensitivity is what makes tissue-culture testing such a valuable screening tool. If questionable results are obtained, the appropriate course of action to be taken is then determined by the study sponsor, study director, laboratory technician, and supervisor of toxicology.

Frequently classes of materials (for example, silicone, polyurethane, and latex) have unique cytotoxic profiles, and thus by knowing the nature of the material being tested it is possible to determine which test would provide the most pertinent data. The agar overlay has been our primary screening method, and we have a very large data base to aid in making the decision as to which method to use.

As new and different types of biomaterials have come into use, we have found that we frequently need more information than the simple pass/fail system of the agar overlay.

In terms of absolute sensitivity, elution of test materials (this includes both the MEM elution test and the low cell density test) provide the most critical cell-culture test for all materials tested. Since the efficiency of the agar overlay is frequently influenced by physical characteristics of the material such as size, weight or, density, these factors must also be considered when making a decision. There are certain materials whose toxicity cannot be determined by agar overlay because of such factors as slow deteriorization, by-products of deteriorization, and so forth that make a low cell density test the more desirable of the two elution tests to use, particularly if the material has not been tested previously or is known to be toxic in very small amounts that possibly could not diffuse through the agar overlay. Other advantages to using an elution method are the ability to determine exact dosages that can be applied once (MEM elution) or repeatedly (low cell density). The extracts can then be included in a battery of physical and chemical tests whose results can be used to accept or reject the material. The low cell density test method can be used to help predict toxicity because of materials degradation while the 24-h MEM elution and the agar overlay frequently cannot.

In spite of the many advantages of using an extract for materials testing, the agar overlay remains our primary screening method because it takes only 24 h. A large number of tests can be run at once, and virtually any type of material can be tested. We are finding that we have a need to be able to do all methods, and this versatility provides us with a wider range of data that in turn make more materials available to the biomedical device field and create a necessity to be constantly evaluating new methods and reevaluating our present systems to be certain that we can and are testing our biomaterials in the most efficient and valuable manner possible.

References

[1] Guess, W. L., Rosenbluth, S. A., Schmidt, B., and Autian, J., *Journal of Pharmaceutical Sciences*, Vol. 54, No. 1, Jan. 1965, pp. 1545–1547.

[2] Autian, J., "The New Fields of Plastic Toxicology—Methods and Results" in Chemical Rubber Company Press, *Incorporated Critical Reviews in Toxicology*, Vol. 2, 1973, pp. 1–40.

[3] Wilsnack, R. E., *Biomaterials Medical Devices and Artificial Organs,* Org:4 (3 & 4), 1976, pp. 235–261.
[4] Wilsnack, R. E., Meyer, F. J., and Smith, J. G., *Biomaterials Medical Devices and Artificial Organs,* Org:1(3), 1973, pp. 543–562.
[5] Wilsnack, R. E., *Biomaterials Medical Devices and Artificial Organs,* Org:4 (3 & 4), 1976, pp. 235–261.
[6] Health Industry Manufacturers Association, "Guideline for Evaluating the Safety of Materials Used in Medical Devices," Health Industry Manufacturers Association Report 78-7, Washington, D.C., June 1978.

Hugh J. Johnson[1] *and Sharon J. Northup*[1]

Tissue-Culture Biocompatibility Testing Program

REFERENCE: Johnson, H. J. and Northup, S. J., **"Tissue-Culture Biocompatibility Testing Program,"** *Cell-Culture Test Methods, ASTM STP 810,* S. A. Brown, Ed., American Society for Testing and Materials, 1983, pp. 25–34.

ABSTRACT: Four methods of in vitro biocompatibility screening using L-929 mouse fibroblast cells are discussed, namely, the fluid medium, agar diffusion, flask dilution, and flask dilution extraction methods. The range of test methods allows toxicity screening of both solid and liquid specimens and has been shown to provide good agreement with and better sensitivity than the United States Pharmacopeia (USP) rabbit muscle implantation assay. Approximately 2000 specimens are subjected to such in vitro evaluations annually at our facility for evaluating stability studies, component changes, formulations, and manufacturing processes. A scale of 0 to 4+ is used to indicate the extent of cellular damage providing a quantitation of toxic response.

KEY WORDS: biocompatibility, tissue culture, materials testing, cell culture

Perhaps the earliest report of a practical, rapid, and sensitive cell-culture method for toxicity screening of plastic materials was presented by Rosenbluth et al.[2] The authors demonstrated that through direct placement of plastic specimens on monolayers of L-929 mouse fibroblast cells a greater number of positive responses was obtained than was elicited with the same materials in rabbit muscle implantation testing. A subsequent paper by Guess et al[3] in the same year showed that incorporation of an agar overlay on the cell monolayers greatly increased both the versatility and usefulness of the basic test, allowing examination of various shaped solids as well as liquid and powder specimens. Also, with the agar method, replicating cultures of L-929 cells proved more sensitive than nonreplicating preparations of primary chick embryo, although an apparent decrease in sensitivity was observed with both cell types when compared to direct placement on monolayers.

[1] Manager and associate director, respectively, Department of Safety Assessment, Travenol Laboratories, Inc., 6301 Lincoln Ave., Morton Grove, Ill. 60053.

[2] Rosenbluth, S. A., Weddington, G. R., Guess, W. L., Autian, J., *Journal of Pharmaceutical Science*, Vol. 54, No. 7, 1965, pp. 156–159.

[3] Guess, W. L., Rosenbluth, S. A., Schmidt, B., Autian, J., *Journal of Pharmaceutical Science*, Vol. 54, No. 10, 1965, pp. 1545–1547.

In this paper, use of L-929 cells in a variety of biocompatibility test systems being used at Travenol Laboratories including the fluid medium, agar diffusion, and other specially adapted cell-culture techniques for rapid screening of large numbers of highly varied materials intended for biomedical use is described. This tissue-culture examination is used as an initial screen or complementary biological evaluation for stability studies, component changes, experimental formulations, changes in manufacturing processes, and testing of final polymer formulations.

Experimental Procedure

Cell-Culture Preparation and General Procedures

The L-929 culture is used with each of four test methods. The American Type Culture Collection (ATCC) is the initial source for the culture that is described as a continuous cell line derived from mouse connective tissue and designation as ATCC CCL 1 NCTC Clone 929. Subcultures are prepared and stored with a cryoprotective agent dimethyl sulfoxide and held in liquid nitrogen for future use. Approximately three subcultures are necessary to produce a cell population of sufficient vitality for use in the test procedures. Optimal growth of the L-929 culture has been obtained through use of minimum essential medium (MEM) with Earle's salts (Flow Laboratories, Rockville, Md.), supplemented with 10 percent calf serum (Flow Laboratories, Rockville, Md.), and containing 100 units and 100 μg/mL, respectively, of penicillin and streptomycin (GIBCO Laboratories, Grand Island, N. Y.).

The L-929 cell line is maintained in continuous cultivation. Transfers are made of confluent monolayers on Monday, Wednesday, and Friday. A cell suspension is prepared for use in the assay by aspirating the growth medium and washing the cells with 0.1% trypsin (GIBCO Laboratories, Grand Island, N. Y.) in calcium-magnesium free Dulbecco's phosphate buffered saline solution (GIBCO Laboratories, Grand Island, N. Y.). The trypsin rinse is aspirated and an additional 1 mL of 0.1% trypsin solution added to each of the 75-cm^2 plastic tissue-culture flasks generally used for monolayer propagation.

The flasks are incubated at 37°C for 5 to 10 min in a maximally humidified atmosphere of 5% carbon dioxide in air to allow time for tryptic activity and suspension of the monolayer growth. When the cells have been totally released from the plastic surface, 10 mL of fresh growth medium is added per flask to halt further enzymatic activity, the yield from all flasks is combined, and the total cell suspension mixed well. An appropriate quantity of growth medium is added to obtain a cell concentration suitable for counting with the hemocytometer. The cell suspension is further diluted with MEM growth medium to a concentration of 1.3×10^5 cells/mL. This suspension is used for preparing test plates for all procedures. Subcultures are prepared by transferring approximately one fifth to one tenth of the cell yield to fresh

75-cm² plastic tissue-culture flasks and bringing the volume to 20 mL with fresh MEM growth medium. For all test methods, the cellular monolayers at the time of exposure to specimens are subconfluent and replicating.

Aseptic and hygienic procedures are used when handling the biological materials. All transfers of cells, media, and reagents are made with mechanical pipet devices. Test plates are prepared on the laboratory bench, but culture transfers and medium preparation are made in a vertical laminar-flow hood.

Solutions added to cells are prewarmed to approximately 37°C, and water used for culture media and reagent preparation is obtained from a deionizer producing water of not less than 1 MΩ resistance.

Specimens for evaluation (exclusive of solutions) are tested in duplicate using both the fluid medium and agar diffusion test methods. As the cellular response with the flask dilution method has been historically uniform, solutions tested with this method are assayed singly rather than in duplicate.

Biocompatibility Screening Methods

Fluid Medium Method

The fluid medium assay determines the toxicity of solid materials to cultured cells in vitro. It is based on the principle that toxic components may leach from the material into the culture medium. Cells at varying distances from the test material will thus be adversely affected by toxic components. Significant toxic results would then be dependent on the extent of cellular damage and its distance from the specimen. Low-density materials that tend to float or move in the culture medium are generally unsuitable for this assay.

Solid specimens are prepared to expose approximately 1 cm² of specimen surface area to the cellular monolayer. When the specimen must be cut to size, the surface that would normally be exposed in use is tested. This is particularly important when cleaning techniques, surface contaminants, or coatings are being evaluated. When the specimen does not have a flat stable test surface, the cut surface may be tested; for example, a cross section of tubing may be cut, and the flat cut surfaces exposed to the monolayer.

After the cell suspension has been diluted to a concentration of 1.3×10^5 cells/mL, 2 mL is pipetted into 35-mm plastic tissue-culture plates. The plates are rotated to distribute the cells and incubated at 37°C for approximately 24 h in a maximally humidified atmosphere of 5% carbon dioxide in air. After incubation and immediately upon removal from the incubator, 1.2 mL of media are withdrawn from each plate, and the specimens gently placed directly upon the monolayer surface. (Medium is not withdrawn when testing highly absorbent material.) One specimen is tested per plate. Four controls are included; two plates containing a nontoxic and two plates containing a toxic specimen. The plates are reincubated at 37°C for approximately 24 h. At the end of this incubation period the specimens are removed carefully with forceps, and growth medium is decanted from the plates. Approximately 0.5 mL of 2% crystal violet in 20% ethanol is rotated over the monolayer, allowed to remain for approximately 1 min, and then rinsed off with water. This technique delineates areas of stained attached cells and clear zones from which dead cells have been detached from the plate surface. After staining, the test material may be replaced within the clear zone so that the difference between the specimen size and the zone perimeter can be quantified. The specimens are judged as passing or failing as in Table 1.

Materials of high density or with unusual shape or surface characteristics may completely or partially denude the cellular monolayer beneath the specimen. Consequently, as degeneration of some or all of the cells beneath the specimen may not reflect toxicity, such specimens receive a rating of "pass." Supplementary test data obtained with concurrently performed agar diffusion procedures, generally a more sensitive indicator, would verify actual specimen toxicity and effectively tend to prevent false negative tests.

A retest is performed if there are meaningful differences between the ratings for the duplicate plates or for the agar diffusion and fluid medium responses, or when the control specimens produce variant results.

Agar Diffusion Method

The agar diffusion assay is used to determine the toxicity of solid, liquid, or powder materials on the assumption that toxic components in the test material will leach through the agar to the cellular monolayer beneath. This assay is well suited for low-density materials (film, paper, and so forth), powders, liquids, and high-density materials that would physically damage the cellular monolayer in the fluid medium method.

Specimens are prepared in the manner described for the fluid medium method. Absorbent materials are soaked briefly in Medium 199 without serum before placement on the agar surface thus obviating cellular damage caused by absorption of water or nutrients by the material. Absorbent materials may also be evaluated using an extraction technique to verify the test results.

TABLE 1—*Subjective grading system for the fluid medium method.*

Rating	Observation
Pass (0, 1+, or 2+)	no discernible cellular changes directly attributable to specimen toxicity
Fail (3+)	cellular malformation and degeneration moderately extended beyond the perimeter of the specimen or moderate decrease in monolayer density or both
Fail (4+)	cellular malformation and degeneration greatly extended beyond the perimeter of the specimen or severe decrease in monolayer density or both

Powder and granular specimens are distributed on the agar surface within an area of approximately 1 to 2 cm^2. Liquids containing organic solvents are tested by soaking filter-paper disks with the specimens, volatilizing the solvent, wetting the dry disk with Medium 199, and placing the disk on the agar. To prepare the assay specimens, 5 mL of cell suspension are pipetted into 60-mm plastic tissue-culture plates. The plates are rotated to distribute the cells and incubated at 37°C for approximately 24 h under 5% carbon dioxide tension and maximal humidity. Immediately upon removal from the incubator, the growth medium is removed from each plate and the cellular monolayer is washed with 2 mL of Dulbecco's phosphate buffered saline. The buffer is removed and 4 mL of Medium 199 with Earle's salts that has been fortified with 1% calf serum, 100 units/mL of penicillin, 100 μg/mL of streptomycin, and 1% agar (Bacto-Agar, Difco Laboratories, Detroit, Mich.) at 45°C, is pipetted gently over the monolayer surface. After solidification of the agar, test, negative (nontoxic plastic) control, and positive (toxic plastic) control specimens are placed on the agar surface. One to three specimens are tested on each plate. Replicate specimens and control specimens are tested on separate plates. After reincubation for approximately 24 h, all plates are examined microscopically at ×40 (bright field) magnification to ascertain the extent of cellular deterioration beneath or surrounding the specimens. Opaque specimens are outlined with pen (on the bottom surface of the plate) and removed from the agar surface before the examination. The specimens are judged as passing or failing the agar diffusion test as in Table 2.

A retest is performed if there are meaningful differences between the duplicate plates for one material or between the fluid medium and agar diffusion responses, or when the controls produce variant responses.

Flask Dilution Method

The flask dilution method is used to determine the toxicity of solutions and is particularly useful for biocompatibility evaluation of parenteral solutions. Solutions to be evaluated should be sterile and warmed to approximately 37°C before use. Temperature-sensitive solutions may be sterilized by

TABLE 2—*Subjective grading system for the agar diffusion method.*

Rating	Observation
Pass	normal cell morphology in proximity to specimen
Fail (1+)	small numbers of degenerated cells beneath specimen area
Fail (2+)	cellular death and degeneration of all cells directly beneath the entire area of the specimen
Fail (3+)	cellular death and degeneration moderately extended beyond the perimeter of the specimen
Fail (4+)	cellular death and degeneration greatly extended beyond the perimeter of the specimen

vacuum or pressure filtration through a sterile filter of 0.2 μm porosity. (Note: the sterile filter should not contain wetting agents that could leach into the test solution.) All other solutions may be sterilized by autoclaving for an appropriate length of time at 121°C. Solutions with relative extremes of osmotic pressure or pH at the usual 0.5-mL test volume are diluted in sterile water to acceptable physiological conditions.

Two mL of cell suspension is pipetted into 35-mm plastic tissue-culture plates. The plates are rotated to distribute the cells and incubated at 37°C for approximately 24 h. After incubation, the medium is removed by aspiration and 1.5 mL of serum-free Medium 199 containing Earle's salts, 100 units/mL of penicillin, and 100 μg/mL of streptomycin is added to each plate. One-half mL of control or test solution is added to single plates. The plates are rotated to mix the added solution in the growth medium. The negative control solution is sterile deionized water. The positive control is a water extract of a toxic plastic material. All plates are incubated at 37°C for approximately 48 h. The toxicity grade and rating of the test solutions are determined microscopically at ×40 magnification (bright field) in relation to the negative control as in Table 3.

Solutions that produce a 1+ or mild 2+ reaction on initial testing are retested. If they receive a retest rating of 0 they are passed. Otherwise, when initial and retest results differ, the higher rating is reported. All specimens are retested if the negative control shows atypical cellular changes or a decrease in normal monolayer density, or if the positive control receives a toxicity rating of less than 3+.

Flask Dilution Extraction Method

This method is essentially an extension of the flask dilution method but allows semiquantitation of test material toxicity. Specimens are generally rubber or other materials inherently toxic to cell cultures. Preparation and

TABLE 3—*The toxicity grade and rating of the test solutions for the flask dilution method.*

Rating	Observation
Pass (0)	confluent monolayer with cells firmly attached and cellular morphology and monolayer density not different from those of the negative control
Fail (1+)	appreciable number of cells malformed or degenerated or mild decrease in monolayer density or both
Fail (2+)	approximately 25% of cells malformed or degenerated or moderate decrease in monolayer density or both
Fail (3+)	approximately 25 to 75% of cells malformed or degenerated or large decrease in monolayer density or both
Fail (4+)	more than 75% of cells malformed or degenerated or moderate decrease in monolayer density or both

evaluation of the test plates and control preparations are as described for the flask dilution method, but serial dilutions of an extract are evaluated.

Extracts may be prepared with various fluids (culture medium, serum, water, balanced salt solutions, and so forth); exposure temperatures and durations, and ratios of material surface or quantity to extracting solution are dependent on the chemical nature and end use of the test material and on the extracting solution. The extract is prepared on the day of the test.

Test materials are generally extracted at a ratio of 6 cm^2 (total specimen surface)/mL to reflect the United States Pharmacopeia (USP) extraction ratio, or 1 g specimen per 10 mL extractant for specimens not amenable to measurement. The extract is added to a series of culture plates containing monolayer preparations in serum-free Medium 199. Twofold serial dilutions representing 0.5 to 0.05 mL of extract are added directly to the test plates. With most preparations, to avoid possible effects caused by relative differences in solution volumes within the test plates, the extraction solutions are brought to the same volume with serum-free Medium 199 or extractant, and 0.5 mL of solution is added to each plate. Further dilutions of the extract are tested if toxicity is encountered at the lowest concentration. Extracting solution controls are also included.

After 48-h incubation of the test plates, the rating for each of the extract dilutions is reported relative to the percentage of original extraction solution added to the plate. To produce the most complete specimen toxicity profile, the extract is diluted to the nontoxic concentration. Specimen toxicity is rated by the scheme used in the flask dilution method (Table 3).

Results and Discussion

Specimens submitted for tissue-culture biocompatibility screening encompass a highly varied group of polymers, metals, discrete chemicals, mixtures, and solutions. Generally, about 2000 specimens are evaluated in a typical year utilizing the four methods described. Most of the synthetic polymers tested in vitro are also evaluated for in vivo compatibility using the USP biological test for plastics. Comparison of the data derived using both test systems offers a convenient means of estimating the efficiency of cell-culture testing in predicting toxicity in the animal.

Three surveys were made of rabbit implant (IM) and tissue-culture (fluid medium and agar diffusion methods) data on more than 1400 materials over an extended testing period. The cumulative results are tabulated in Table 4. The percentage of total agreement (pass/pass and fail/fail) between the tissue culture and IM testing was 82, 96, and 92% in the three surveys. The data also indicate that the implant method is less sensitive than the tissue-culture method, tending to pass materials failed using the in vitro method.

TABLE 4—*Composite of in-house test data comparing relative sensitivity of in vitro and in vivo biocompatibility evaluation methods.*

		Data Correlation, percentage			
Survey	Specimens Evaluated	TC[a] Pass and IM[b] Pass	TC Pass and IM Fail	TC Fail and IM Pass	TC Fail and IM Fail
1[c]	793	72	12	6	10
2	470	96	0	4	0
3	217	92	0	8	0

[a] TC = tissue culture test.
[b] IM = rabbit muscle implantation test.
[c] This early survey involved agar diffusion results of 1+ that were then classified as pass, and 2+ fluid medium results that were considered pass or fail depending on specimen characteristics. In addition, observation of the IM reactions after 3-day, rather than 7-day, implant increased the opportunity for interpreting tissue trauma as specimen toxicity.

A group of 290 specimens consisting primarily of natural and synthetic rubber materials and a small number of polyvinyl chloride (PVC) polymers was evaluated using National Formulary (NF) extraction and the fluid medium and agar diffusion test methods (Table 5). Agreement between the NF method using two extracting solutions and solid specimen test methods was 74.8% (specimen toxicity and nontoxicity detected by both methods). With 11.1% of the specimens the relatively stressful NF 121°C extraction for 60 min was apparently more effective than the gentle aqueous/serum solid extractions performed at 37°C in extracting polar (saline) and nonpolar (oil) leachables. The opposite was true with 13.8% of the specimens. In many in-

TABLE 5—*Comparison of national formulary extraction and solid specimen methods of biocompatibility evaluation using plastic and elastomeric materials.*

Number of Specimens (% of Total)	Solid[a]	NF[b] Saline Extract	NF[b,c] Oil Extract
40 (13.8)	+[d]	+	+
47 (16.2)	+	+	−[e]
33 (11.4)	+	−	+
40 (13.8)	+	−	−
6 (2.1)	−	+	+
17 (5.9)	−	+	−
9 (3.1)	−	−	+
98 (33.4)	−	−	−
Total 290 (100)			

[a] Solid = tested using fluid medium and agar diffusion methods.
[b] NF = national formulary; 1.25 cm^2 specimen/mL extracted with 0.9% saline or cottonseed oil at 121°C for 60 min; flask dilution method used.
[c] Cottonseed oil; some sesame and soybean also used.
[d] + = toxic reaction.
[e] − = nonreactive.

stances with the saline or oil extract there was a maximal 4+ response while the solid specimen showed the test substance to be compatible and vice versa.

Conclusions

The L-929 continuous cell line offers several advantages for use in biocompatibility screening programs. The cell-line multiplication rate is quite rapid in relatively simple growth media. At the cell concentrations transferred and subculture schedules used, optimal monolayer vitality is maintained that, in turn, has promoted uniformity and stability of cellular response. The outline of individual cells in the monolayer is distinct as the cells tend to grow in loose contact with one another. These characteristics simplify discerning changes in cell morphology, which is an essential criterion for rating specimen toxicity. The addition of penicillin and streptomycin to the culture medium eliminates microbial contamination without an apparent compromise of sensitivity.

The fluid medium method is the least complicated of the four discussed. Fixation of the monolayer with crystal violet-alcohol allows review of the plates at a later date, better defines the cellular response, and reduces the number of microscopic observations necessary. Quantitation of toxicity (on a scale of 0 to 4+) has proved a useful and reproducible adjunct in evaluations, although for the fluid medium method the differentiation is less clear with classifications of 1+ and 2+ being judged nontoxic because of cell trauma.

The chemistry of the agar overlay theoretically has the potential to impede or prevent diffusion of certain leachates to the cellular monolayer beneath. Working with test materials containing essentially water insoluble components and contaminants, we have had better sensitivity with the agar overlay method than with the fluid medium assay. This may in part be because 1+ and 2+ reactions are not considered toxic with the latter method. On the other hand, small changes in cellular morphology can be readily discerned microscopically with the agar overlay; variability between duplicate plates and among initial and retest assays occurs infrequently with this method. In addition, based on both historical familiarity with the method and statistical considerations, the retest rating system for the agar diffusion method has been refined to increase accuracy of the toxicity evaluation.

A combination of the fluid medium and agar diffusion methods has been shown to exceed the sensitivity of the widely used USP rabbit implant method (Table 4). The comparative data in Table 5 indicate that within a group of complex material types, differences in toxic/nontoxic specimen ratings can occur in about one out of four materials evaluated using solid polymer and polymer extraction testing. Consequently, a combination of solid and extract test methods would appear to offer increased sensitivity; however, this may be at the expense of a positive correlation with in vivo compatibility.

Based on our experience and the data presented, the tissue-culture biocompatibility test procedures described offer uncomplicated, rapid, and sensitive means of biomaterial toxicological examination. The sensitivity of the cell culture in vitro system exceeds that of the USP rabbit implantation method. This indicates that the in vitro method may be a reliable substitute for the in vivo test system.

Charles R. Horres,[1] *Hamilton G. Mears,*[2]
Roger E. Wilsnack,[2] *and George E. Moore*[3]

Evaluation of Materials Used for the Fabrication of Medical Devices: Cytotoxicity Testing

REFERENCE: Horres, C. R., Mears, H. G., Wilsnack, R. E., and Moore, G. E., **"Evaluation of Materials Used for the Fabrication of Medical Devices: Cytotoxicity Testing,"** *Cell-Culture Test Methods, ASTM STP 810,* S. A. Brown, Ed., American Society for Testing and Materials, 1983, pp. 35–50.

ABSTRACT: Becton Dickinson has been involved in cell-culture testing of medical devices for the past ten years using WI-38 cells as a human origin in vitro target system. Two modes of tissue-culture testing are used at Becton Dickinson: cytotoxicity score and inhibition of cell growth. To determine score, cell monolayers are scored microscopically as to the degree of morphologically discernable cytotoxicity after incubation for 24 and 72 h with culture media used to eluate test materials. Inhibition of cell growth is determined by culturing cells for 24 h in test material eluate then trypsinizing and reseeding using the elution fluids as growth menstruum. The plates are incubated, and when negative controls indicate a 100% monolayer, the cell monolayers are fixed and stained. The cells are scored microscopically as to the percent of growth inhibition in relation to the negative control. This system has provided a sensitive, reproducible, and short duration evaluation mechanism for biomaterials.

KEY WORDS: cell culture, cytotoxicity, cell growth

Since the introduction of tissue culture as an in vitro method for screening medical materials and devices in 1960, the use of tissue-culture systems has expanded considerably [*1*].[4] In view of the goal of biomaterials biocompatibility testing of forecasting the toxicity of materials in clinical applications, tests based on the use of nonmalignant cells of human origin combined with a physiological extraction media for material extraction should provide more relevant results than toxicity tests using cell lines from other species and nonphysiological extraction fluids. Becton Dickinson has pioneered in the use of human female embryonic lung fibroblasts for testing culture-media

[1] Department of Physiology, Duke University, Durham, N. C. 27706.

[2] Research technologist and director, respectively, Becton Dickinson Research Center, Post Office Box 12016, Research Triangle Park, N. C. 27709.

[3] Lab director, Biosearch, Philadelphia, Pa. 19056.

[4] The italic numbers in brackets refer to the list of references appended to this paper.

extracts of medical device materials [2]. Two forms of human cell-culture tests have been used at the Becton Dickinson Research Center for routine biocompatibility testing. These include a cytotoxicity test of culture-media extracts of material placed on confluent monolayers of human fibroblasts for periods of 24 and 72 h and a growth inhibition test in which test material culture media eluate is used as growth media before and after reseeding human fibroblast cells in order to monitor effects on cell growth. To generate meaningful and reproducible results with a tissue-culture system, careful attention must be given to selection of extraction methodologies, interpretation of cell response, and monitoring of the tissue-culture systems.

In this review, the procedures used at Becton Dickinson to obtain consistent results in material evaluation with cell culture are presented. In addition, the concept of surface area dosimetry and its potential application to medical device design is introduced.

Methods

The primary purpose of this review is to document the methods and procedures involved in tissue-culture testing for biomaterial evaluation at Becton Dickinson. These not only include procedures involved with culture preparation but also methods for specimen preparation, test protocols, and quality control. Because of the sensitivity of the culture system, minor methodological details can make the difference between a successful program and one producing inconsistent results. The methods and materials outlined below are those that have worked successfully at Becton Dickinson.

Tissue-Culture Incubators

Forced convection carbon dioxide incubators (Model 3028, Forma Scientific, Marietta, Ohio 45750) are used for cell-culture and media testing. The incubators are maintained at 37 ± 0.5°C, 90 to 95% relative humidity, and 5 ± 0.5% carbon dioxide air. Water supplied to the incubators is deionized, and once monthly the incubators are drained and disinfected with 70% ethyl alcohol/water. Temperature is continuously recorded and carbon dioxide checks are performed daily.

Sterile Preparation Area

Laminar flow hoods (SterilGard Model VBM-600, Baker Co., Sanford, Me. 04073) are used for all culture manipulations requiring aseptic techniques. These operations include media preparation, cell plating, eluate addition, and subculturing. As an added precaution, materials suspected of being contaminated or under quarantine are handled in a separate hood. Hoods are recertified annually and maintained on a periodic basis.

Culture Vessels

Depending on the particular aspect of the culture procedure and volume of eluate available for testing, a variety of sizes of polystyrene culture dishes, plates, and flasks are used. These include 60-mm petri dishes (#3002 tissue culture dish, Falcon, Oxnard, Calif. 93030), 100-mm petri dishes (#3003 Falcon), 24 well plates (Multiwell #3047 Falcon), and 96 well plates (Microtest II, #3040 Falcon). Tissue-culture flasks include 25-cm^2 (T-25) growth surface (#3012 tissue-culture flask), 75-cm^2 (T-75) flasks (#3024 Falcon), and 150-cm^2 (T-150) flasks (#25120-150 Corning Glass Works, Corning, N. Y. 14830).

Culture-Media Preparation

Culture medium used for tissue-culture testing is prepared in 4-L batches, and lot numbers of all medium components are recorded. The base formulation is minimum essential media (MEM) with Earle's salts (GIBCO #320-1380, Grand Island Biological Co., Long Island, N. Y. 14072) and is supplemented with 1% volume/volume (V/V) glutamine (GIBCO #320-5030), 1% V/V sodium pyruvate (GIBCO #320-1140), and 1% V/V nonessential amino acids (GIBCO #320-1140). Antibiotic and antifungal agents include 0.5% V/V Fungizone® (GIBCO #600-5295), 1% V/V penicillin-streptamycin solution (GIBCO #5145), and 0.1% V/V gentamycin sulfate (Garamycin, Schering Corporation, Kenilworth, N. J. 07033).

Water for reconstitution of reagents is United States Pharmacopeia (USP) sterile water for injection (#2A0302, Travenol Laboratories, Deerfield, Ill. 60015). Media pH is adjusted to pH 7.2 with sterile 1 *N* sodium hydroxide before adding 10% fetal bovine serum. Media is stored for no longer than one month at 4°C, and if older than two weeks, an additional 1% V/V L-glutamine is added before use. Each media batch is tested for sterility by innoculating trypticase soy broth and thioglycolate media and checking for growth after seven days of incubation at room temperature and 37°C.

Serum Screening

Serum is screened routinely before use in tissue-culture media. Specimens representing several lots and manufacturers are obtained for evaluation. The serum screening media consists of 15% V/V of the test fetal bovine serum (Flow Labs, McLean, Va. 22102; Sterile Systems, Inc., Logan, Utah 84321), 2% V/V L-glutamine, 1% V/V sodium pyruvate, 1% V/V nonessential amino acids, 0.1% V/V gentamycin, 1% V/V penicillin-streptamycin, and 0.5% V/V Fungizone. Because of increased cloning efficiency over WI-38, the MRC-5 cell line is used to screen serum. Media is conditioned by exposure to monolayers of MRC-5 cells (American Type Culture Collection CCL 171, Rockville Md. 20850) for 24 to 72 h using media volume of 50 mL to cell area of 150 cm^2, filtered (0.45 μm), and stored. Cell suspensions are prepared using 50% conditioned test media and 50% fresh test media. Using a MRC-5 concentration of 30 cells/mL, 60-mm dishes are seeded with 6.5-mL of suspension (about 200 cells) for each lot of test serum. Relative plating efficiencies are compared (number of colonies formed/number of cells seeded) by counting colonies on an automated colony counter (Model C111, New Brunswick Scientific Co., New Brunswick, N. J. 08901). In addition, average colony size is observed macroscopically and used along with plating efficiencies to evaluate serum lots. Serum giving satisfactory response is purchased in 100 L lots and frozen at −70°C for storage.

Cell Line

The cell line used for cytotoxicity and growth inhibition is WI-38 (diploid) (American Type Culture Collection CCL 75, Rockville, Md. 20850). Cells are obtained frozen, thawed in a 37°C water bath, and then diluted at a ratio 1:10 in culture media. Cells are subcultured until six to eight T-150 flasks are confluent. The flasks are washed with calcium, magnesium free Hank's balanced salts (GIBCO #310-4170), exposed to 5-mL trypsin solution (GIBCO #610-5050) for 1 min at 37°C, the trypsin solution decanted, and the flasks incubated at 37°C until cells have rounded (8 to 30 min). Cells are recovered in 10 mL of culture media, centrifuged, and reconstituted to 2×10^6 mL in medium containing 20% V/V fetal bovine serum, 1% V/V L-glutamine, 1% V/V penicillin-streptamycin, and 5.5% V/V dimethylsulfoxide. One-mL aliquots are frozen in plastic ampules (NUNC, Vangard International, Neptune, N. J. 07753) in a controller-freezer (Model 801 Cryo-Med, 49659 Leona Dr., Mt. Clemens, Mich. 48045) at a freeze rate of 2°C/min. One ampule is reconstituted immediately and cultured. If the culture time to reach confluency in a T-75 flask exceeds five days, the lot of cells is rejected. An additional T-75 flask of cells is grown to confluency and sent to mycoplasma assay after freezing the cells in the flask (Flow Labs, 7655 Springhouse Rd., McLean, Va. 22102). Mycoplasma positive lots are rejected. Cells are not passaged more than four times after reconstitution.

Preparation of Culture for Cytotoxicity

To prepare confluent monolayers for cytotoxicity testing, one ampule of WI-38 cells is thawed rapidly at 37°C and reconstituted in 20 mL of complete media and incubated in a T-75 flask with media exchanges every 24 or 48 h until confluent. After trypsinization, the culture is then subcultured at 1:2 in two T-150 flasks with 80-mL complete medium. When confluent, one of the flasks is used for seeding test plates, and the other continued in subculture production until four passages have been reached. Cells used for seeding are trysinized and checked for viability by trypan blue exclusion. Seeding density is adjusted to 5 × 10^4-mL viable cells in complete medium before aliquoting to culture vessels. In 60-mm dishes 5-mL of cell suspension is used, and 10 mL of cell suspension is used in 100-mm dishes. The 24 well plates receive 1 mL/well and 96 well plates receive 0.2 mL/well. Plates are incubated at 37 ± 0.5°C, 5 ± 0.5% carbon dioxide air, 90 to 95% relative humidity until confluent before use in cytotoxicity tests. Cells are refed by addition of ½ of the original media volume 24 h before test eluate is applied.

Specimen Preparation

Specimens to be tested are weighed carefully and placed in sterile borosilicate glass bottles (#10-158-1 Teflon® lined cap, Wheaton Scientific, Millville, N. J. 08332), and 1:5 weight/volume ratio of complete culture medium added. Specimens are incubated with the elution fluid for 24 h at 37 ± 0.5°C without agitation. If the eluate appears turbid it is centrifuged at 1475 *g* for 10 min to obtain clear supernate; otherwise, it is used directly for testing. Each previously prepared dish or well is examined for confluency and characteristic swirling growth pattern before use. Growth medium is suctioned off, and an equal amount of prewarmed (37°C) test eluate added.

Two wells or dishes are used for each specimen. A positive control eluate at 1:8, 1:16, and 1:32 dilutions and a known negative control are run with every batch of test eluates. Cultures are maintained at 37 ± 0.5°C, 5 ± 0.5% carbon dioxide air, 90 to 95% relative humidity for 24 or 72 h.

We have designed our test, through incorporation of antibiotics at noncytotoxic levels, to avoid the necessity of sterilization of each specimen, and even though extracted specimens are not always sterile, contamination of cultures has been rare. Care is taken in handling the specimens to minimize contamination.

Cytotoxicity Score

Test plates or dishes are decanted and rinsed with phosphate buffered saline, pH 7.0, then fixed in absolute methanol for 15 min and air dried. Plates are stained with filtered Harleco hematoxylin stain, Harris formula (#S7735-3, American Scientific Products, McGaw Park, Ill. 60085) for 1 h, rinsed with water, and dried. Plates are reexamined on an inverted tissue-culture scope at ×40 (Model 1810 American Optical, Buffalo, N. Y. 14240) and compared with negative controls using the criteria of morphological appearance, cell density, and distribution. Scoring is on a basis of 1 to 4 with 4 indicating maximal cytotoxic response. Positive control plates are also rated and reported as a cytotoxic index.

Human-Cell-Growth Inhibition Test

This test is conducted in 24 well plates and begins with a 24-h incubation with test eluate on confluent cultures as in the cytotoxicity test previously described. After 24 h, cells are trypsinized and reseeded in different wells with prewarmed test eluate. When the negative control is confluent, the cells are fixed and stained as previously described. Results are scored relative to negative controls in increments of ten and presented as percent inhibition relative to controls. Eluates producing greater than 20% inhibition are considered positive. Positive controls are also scored and recorded.

Results

Tissue-culture cytotoxicity tests and growth inhibition tests are used in many phases of medical device manufacture at Becton Dickinson. These include material validation for new product designs, new materials for existing product lines, batch testing of raw materials, and finished product release tests. The number of tests run in a recent series by type and response are given in Table 1. Most of these tests were run in 24 well plates. The 72-h test and growth inhibition tests are used most often on materials found in previous screening to produce delayed responses or suspected from other data as potentially cytotoxic, which accounts for the increased percentage of positive results in these tests. Delayed responses are often present with polyvinylchloride specimens. When positive eluates are identified with new materials that are at maximum cytotoxicity, the procedure is repeated with twofold dilutions up to 1:64 in order to determine the cytotoxic end point. Of the types of materials tested, polyvinylchlorides and rubber compounds produce most of the positive response.

Determination of cytotoxic score requires considerable experience in microscopic evaluation of the WI-38 cell morphology and growth patterns. Some of the significant characteristics examined for signs of cytotoxicity will be described. A healthy WI-38 monolayer is a confluent pattern of swirls, similar to a fingerprint, and at ×40 the monolayer will seem smooth and uninterrupted. The cells are long and slim and are bounded on all sides by other cells. The nucleus is not prominent but can be seen, and there are very few rounded cells (Fig. 1).

Several conditions in healthy cultures can be mistaken for mild toxic responses by the inexperienced examiner. For example, if the cell density is

TABLE 1—*Tissue-culture tests conducted during a recent test series.*

Test Type	Number	Positive, %
24-h cytotoxicity	862	6
72-h cytotoxicity	13	38
Growth inhibition	21	29

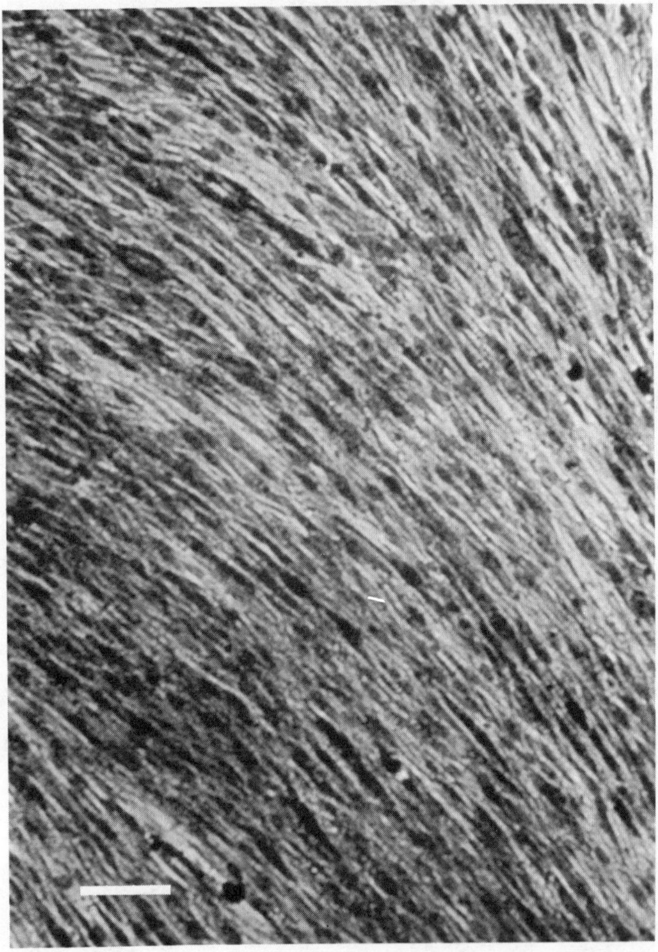

FIG. 1—*Negative control for cytotoxicity test. Calibration scale* = 50 μm.

very great or the monolayer is old (2 to 5 days of confluency), which means that nutritional stress is likely, the cells may tend to pull away from the exterior walls of the well. This could be mistaken for a toxic response. Also under these conditions, if the centers of two swirling patterns are close together the cells in between can be pulled apart, opening a hole in between two areas where cell density is very great. This is another occurrence that could be mistaken for a toxic response.

One of the more common and easily identified toxic responses is one that destroys all cells, in which case the wells will be empty, or there will be fragments and debris only. Frequently toxicity is evidenced by cell sloughing,

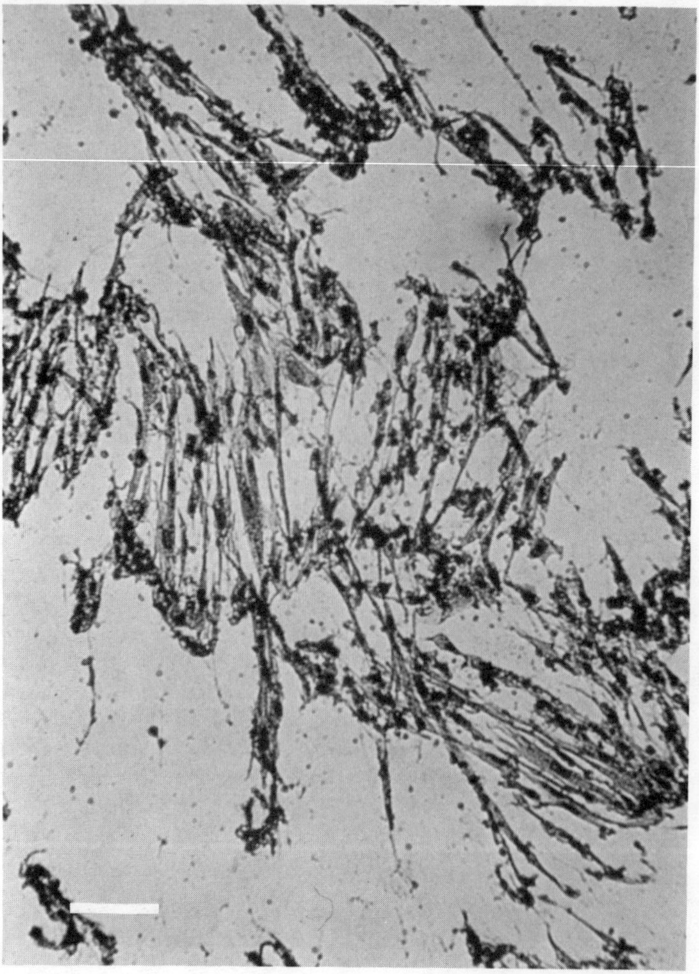

FIG. 2—*Cytotoxic Score 4. Calibration scale* = 50 μm.

holes in the monolayer, and considerable debris and granulation (Fig. 2). In a case where the response is less severe, the culture is inspected for holes in the monolayer (Fig. 3). If holes are on the edge of the wells or next to swirl patterns, they must be confirmed by appearance in duplicate wells and checked against the negative controls to be sure they are not the effect of cell density or monolayer age described previously. Cells with the cytoplasm missing and "strings" of debris present next to holes in the monolayer are strong evidence of toxicity. Holes in the monolayer that are elongated, in the general shape of fibroblasts, and are distributed evenly through the monolayer are also a positive sign of toxicity (Figs. 4 and 5).

Another type of toxic response results in short, stubby cells with a pro-

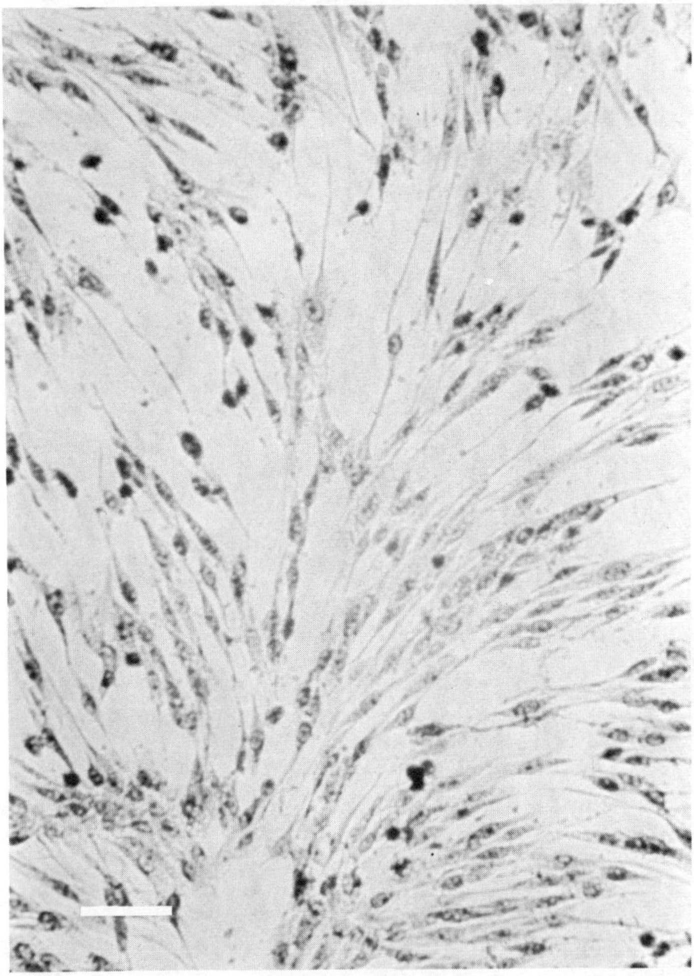

FIG. 3—*Cytotoxic Score 3. Calibration scale* = 50 μm.

nounced nucleus. A large percentage of cells may be affected, but the monolayer can still appear confluent in appearance. This response may or may not be associated with a granular appearance, and there is usually not a lot of debris in this case (Fig. 6).

A score of 1 is given when, compared to the negative control, there is any evidence of what has been described previously. If 25 to 50% of the cells have been affected it is scored as 2, 50 to 75% is scored 3, and greater than 75% is scored 4. The incidence of false positive results based on retest statistics has been very low and the interest coefficient of variation on positive controls is approximately 6% based on cytotoxic end point titration.

In the growth inhibition test the percentage of cells that can reattach after

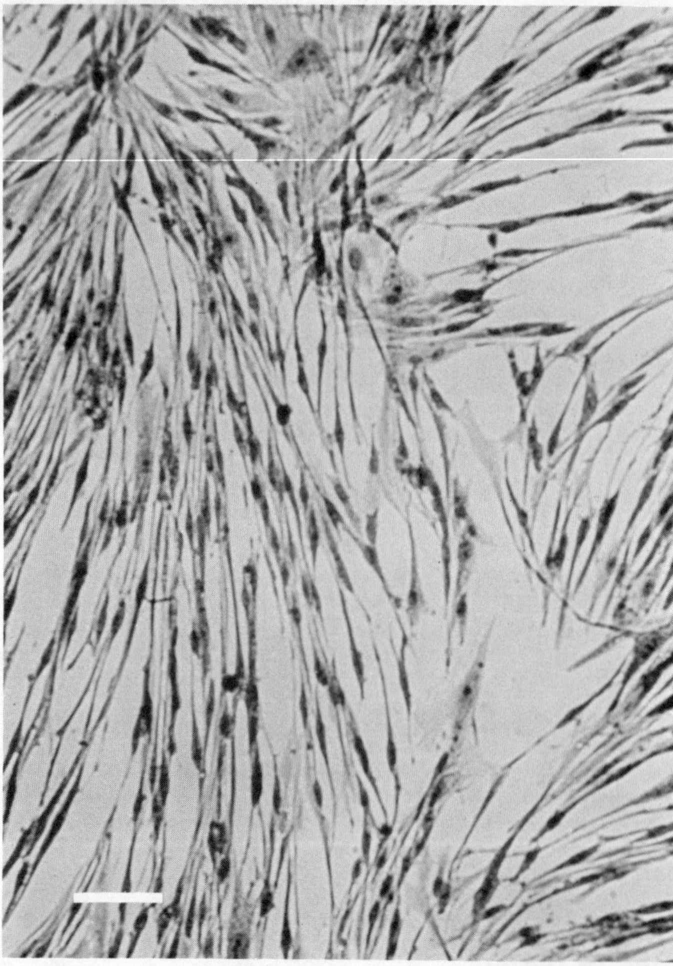

FIG. 4—*Cytotoxic Score 2. Calibration scale* = 50 μm.

trypsinization is compared to that of the negative control and forms the primary criteria for judging growth inhibition. This must be considered with other evidence of toxicity, because it is possible to reseed wells at a different density than the control. Percent inhibition is scored after microscopic examination (×40) in increments of 10% with 20% or greater inhibition considered positive. In some instances cells will be completely unable to attach in the presence of a test eluate, and the well will be empty when it is fixed and stained.

When growth inhibition specimens are trypsinized and moved into new wells after the first 24 h of incubation, the new wells are seeded with a large number of cells. Because of this the monolayers that develop often have a

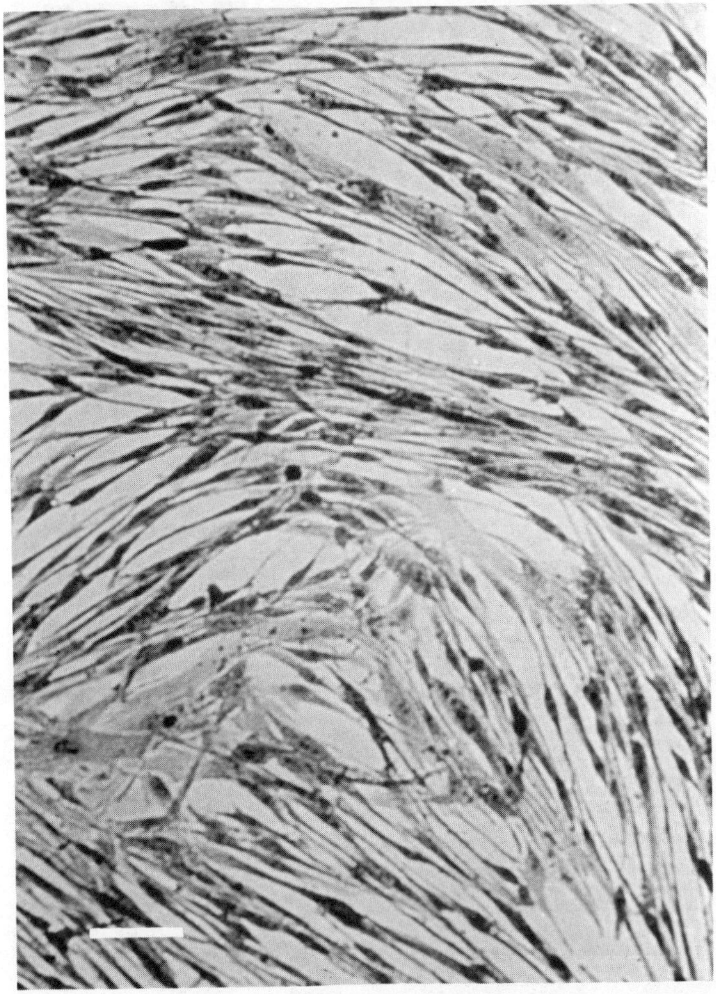

FIG. 5—*Cytotoxic Score 1. Calibration scale* = 50 μm.

very high concentration of cells in the center of the well, with the perimeter being less populated (Fig. 7). Physical tension will pull cells away from the side walls of the wells, leaving gaps in the monolayer that could be mistaken for a toxic response. For this reason, it is important to judge the center portion of each well more critically than the perimeter when comparing the growth inhibition specimen with its control.

Discussion

Becton Dickinson has obtained a decade of experience in the use of cell cultures for toxicity testing and has established that the WI-38 cell-culture

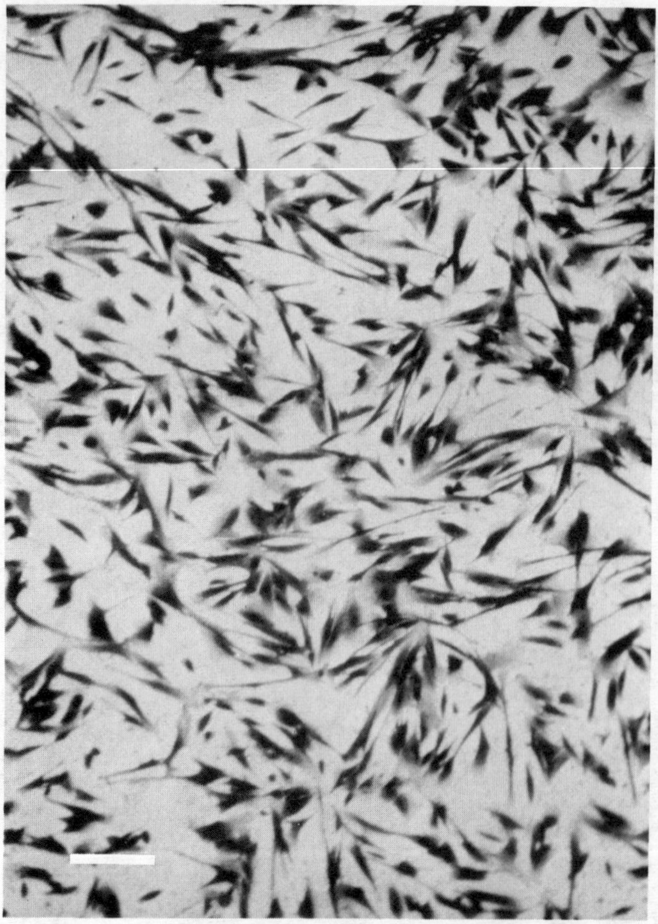

FIG. 6—*Cytotoxic response producing short cells. Calibration scale* = 50 μm.

test provides a sensitive indication of cytotoxicity. In addition, the WI-38 cell-culture test is reproducible and is a test of relatively short duration. The rationale for these tests is twofold. First, the WI-38 cell line is human in origin, noncancerous, and has a diploid chromosonal configuration that approximates the biology of the clinical application of the test material [*3*]. Second, the medium used to elute materials is the most appropriate analog of extravascular fluid because serum components of the media are capable of solubilizing molecules not directly soluble in saline. Our earlier work established that culture medium provides a better correlation between culture response and in vivo response than other extraction media [*4*]. Similar results have been reported by other biomaterials test groups with this cell line [*5*]. An additional advantage of this cell line is the considerable amount of in-

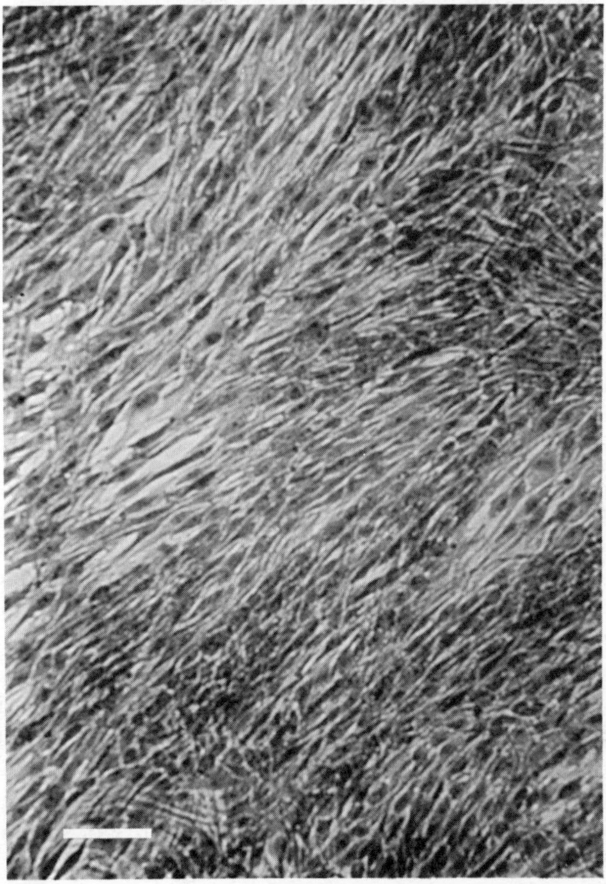

FIG. 7—*Growth inhibition negative control. Calibration scale* = 50 μm.

formation available on the morphological, physiological, and biochemical characterization of the WI-38 cells [*6–13*]. When stocks of WI-38 become depleted, the MRC-5 and IMR-90 lines of human embryonic lung fibroblasts would be excellent alternatives [*14,15*]

In our experience we have found it absolutely essential to carefully monitor the culture system. This includes appropriate biological monitoring, calibration of instruments, and sanitation and scheduled maintenance of culture incubators and laminar flow hoods. Since serum can vary considerably from lot to lot, we view serum screening as mandatory, and because of the insidious nature of *Mycoplasma* contamination we feel that all incoming cell lines should be tested before use in the evaluation of biomaterial [*16*]. In addition, we do not continue passage of cells beyond four doublings from our initial freeze, approximately 20 to 24 from tissue of origin at use, which helps keep the growth rate of the cultures within a useful range [*15*]. Training and attention to detail by the tissue-culture technician are also important factors in the consistency and reliability of test results.

The greatest source of variation in cell-culture tests is the subjective assessment of cytotoxicity. As with most histopathological evaluations, there is no substitute for the well trained and experienced histotechnologist. We use several techniques to improve interpretation of cytotoxic responses and confidence in test results. These include parallel determination of known positive and negative materials as controls. As a further indication of culture-system sensitivity, we routinely determine the maximum dilution of a known positive control that produces a response and report the values as a cytotoxic index. We found it possible to save fixed plates for several years with good results, which permits training of the histotechnologists and reverification of test results.

Cell-culture testing at Becton Dickinson has involved testing on a weight to extractant ratio originally used by Brewer and Bryant [*17*], or at special request, under USP recommended dosimetry guidelines for biological tests on plastics. The USP recommends 60 cm^2/20 mL for specimens equal to or greater than 0.5 mm in thickness and 120 cm^2/20 mL for specimens thinner than 0.5 mm [*18*]. These doses do not reflect the exposure anticipated for medical device use or provide a meaningful basis to evaluate material or products with unique surface treatments. In addition to the difficulty of relating exposure and surface treatment to extraction protocols, a single-point response does not provide sufficient information to formulate interpretable results when extrapolating to specific device patient exposure. Although we routinely determine a cytotoxic index by dilution of specimen eluates of a positive material, this does not allow for the possibility of saturation of the eluate by a poorly soluble cytotoxin.

We have been developing a concept for testing based on surface area dosimetry that hopefully will provide a more quantitative approach to toxicity testing. Since many test materials are available in surface to volume ratios that allow extraction rations up to 30 to 60 cm^2/mL, especially if gentle tumbling is used in the extraction process, it is possible to extract over a wide range of multiples of surface area to generate a series of extraction ratios. From cell-culture tests on such a series of specimens, a dose response curve can be generated (Fig. 8).

By relating the level of exposure that is nontoxic in culture to acceptable safety factors, it may be possible to construct nomograms that would allow the product designer to either evaluate the suitability of a given material for specific device fabrication or, where a particular material has other desirable characteristics, to design or recommend use of a device within a safety factor for exposure based on surface area.

In summary, our experience has established the use of the WI-38 cell-

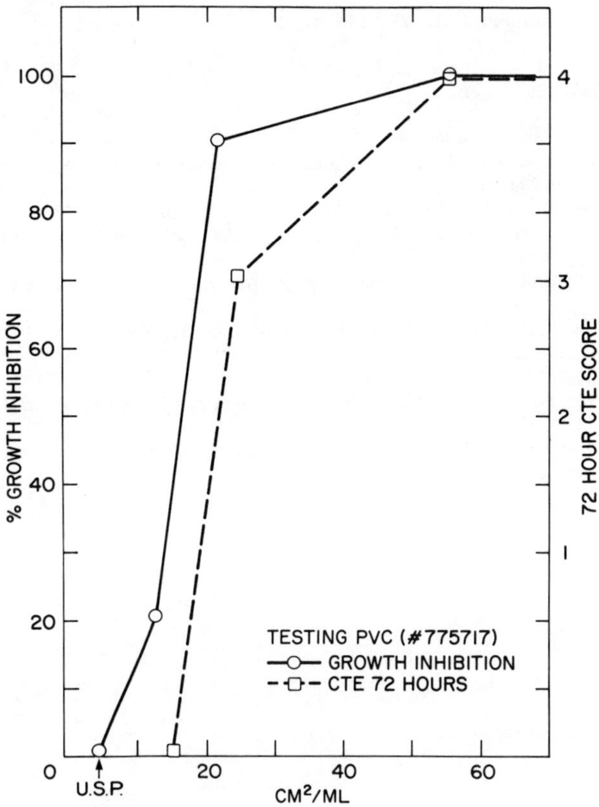

FIG. 8—*Surface area dose response curve.*

culture system as an important aspect of biomaterials evaluation. We anticipate that by developing more realistic protocols for specimen extraction, the applications of this approach can be extended.

References

[1] Cruickshank, C. N. D., Hopper, C., Lewis, H. B. M., and MacDougall, J. D. B., *Journal of Clinical Pathology,* Vol. 13, No. 1, Jan. 1960, pp. 42–50.

[2] Wilsnack, R. E., Meyer, F. J., and Smith, J. G., *Biomaterials, Medical Devices, and Artificial Organs,* Vol. 1, No. 3, 1973, pp. 543–562.

[3] Hayflick, L., *American Review of Respiratory Disease,* Vol. 88, No. 3, Part 2, Sept. 1963, pp. 387–393.

[4] Wilsnack, R. E., *Biomaterials, Medical Devices, and Artificial Organs,* Vol. 4, No. 3/4, 1976, pp. 235–261.

[5] Gill, M., "Direct Cell Contact Screening for Materials and Devices," *MD&DI,* March 1982, pp. 72, 74–76.

[6] Packer, L. and Smith, J. R., *Proceedings of the National Academy of Science,* Vol. 71, No. 12, Dec. 1974, pp. 4763–4767.

[7] Maizel, A., Nicolina, C., and Baserga, R., *Journal Cell Physiology,* Vol. 86, No. 1, Aug. 1975, pp. 71–82.

[8] Ham, R. G., Hammon, S. L., and Miller, L. L., *In Vitro,* Vol. 13, No. 1, Jan. 1977, pp. 1–10.

[9] Wolosewick, J. J. and Porter, K. R., *American Journal of Anatomy,* Vol. 149, No. 2, June 1977, pp. 197–226.

[10] Rawles, J. W., Jr., and Collins, J. M., *The Journal of Biological Chemistry,* Vol. 252, No. 14, July 1977, pp. 4762–4766.

[11] Crusberg, T. C., Hoskins, B. B., and Widdus, R., *Experimental Cell Research,* Vol. 118, No. 1, Jan. 1979, pp. 39–41.

[12] Litwin, J., *Journal of Cell Science,* Vol. 40, Dec. 1979, pp. 281–291.

[13] Hazelton, B., Mitchell, B., and Tupper, J., *Journal of Cell Biology,* Vol. 83, No. 2, Part 1, Nov. 1979, pp. 487–498.

[14] Jacobs, J. P., Jones, C. M., and Baille, J. P., *Nature,* Vol. 227, No. 254, July 1970, pp. 168–170.

[15] Houghton, B. A. and Stidworthy, G. H. *In Vitro,* Vol, 15, No. 9, Sept. 1979, pp. 697–702.

[16] Sethi, K. K. and Teschner, M., *Klinische Wochenschrift,* Vol. 50, No. 5, March 1972, pp. 226–233.

[17] Brewer, J. H. and Bryant, H. H., *Journal of the American Pharmaceutical Association,* Vol. 49, No. 10, Oct. 1960, pp. 652–656.

[18] Guidelines for Biological Tests on Plastics, in *The United States Pharmacopeia XX Revision,* Mack Publishing Co., Easton, Pa., 1980, p. 951.

Elwood O. Dillingham[1]

Primary Acute Toxicity Screen for Biomaterials: Rationale, In Vitro/In Vivo Relationship and Interlaboratory Performance

REFERENCE: Dillingham, E. O., **"Primary Acute Toxicity Screen for Biomaterials: Rationale, In Vitro/In Vivo Relationship and Interlaboratory Performance,"** *Cell-Culture Test Methods, ASTM STP 810,* S. A. Brown, Ed., American Society for Testing and Materials, 1983, pp. 51–70.

ABSTRACT: Correlation analysis of in vitro and in vivo toxic response to biomaterials, based on tests done directly on the materials and the material extracts (water, saline, polyethylene glycol-400, and cottonseed oil), demonstrated significant in vitro/in vivo correlations. In vitro tests that most strongly correlated with in vivo tests were inhibition of cell growth, hemolysis, and agar overlay tissue culture (zone index for test directly on the material). In vivo tests that most strongly correlated with in vitro tests were intradermal irritation (cottonseed oil and polyethylene glycol-400 extracts), systemic toxicity (cottonseed oil and polyethylene glycol-400 extracts of high toxicity materials), and intramuscular implantation (gross rating of the material implant). In vitro and in vivo toxicity ratings were correlated with histopathologic tissue response to material implants (necrosis [coagulative necrosis and overall degree of necrosis], inflammatory cellular infiltrates [polymorphonuclear leukocytes, macrophages, and giant cells], and fibroendothelial proliferation), although the correlations were lower than for the primary screen tests per se.

Interlaboratory performance of the primary screen indicated a useful degree of discrimination of the toxicity of different materials; all laboratories ranking the test materials in the same order with respect to toxicity. Significant differences in the absolute toxicity ratings among laboratories indicated that laboratory performance with respect to standard reference materials will be required for meaningful comparison of toxicity ratings from different laboratories.

KEY WORDS: biomaterials, acute toxicity, extracts, protocols, in vitro/in vivo correlations, interlaboratory performance, testing rationale

Contemporary developments in cell-culture methodology coupled with the current projection of "in vitro imperative," derived from technological, economic, and societal needs, provided the context of this symposium. In vitro assays have evolved as an essential component of the primary acute

[1] Associate professor and head of Cellular Toxicology, Materials Science Toxicology Laboratories, Memphis, Tenn. 38163.

toxicity screen for biomaterials. The utilization of in vitro assays by different investigators differs markedly even though the objective of prediction of higher order (clinical or human) response is a common denominator. Superimposed on the diversity of in vitro technology is the diversity of physical and chemical properties of biomaterials or biomedical devices, which limit the utility or validity of single assays. The end points for the assessment of in vitro assays are often complex and the higher order response to be predicted even more complex. An increase in the precision or objective reproducibility of an in vitro assay can be usually achieved but at the expense of technological difficulty and a narrowing of the scope and applicability of the assay. Faced with the need for assessment of diverse biomaterials and chemical agents associated with such materials, there is a need for assessment of predictive values of technically simple, broadly applicable tests. This demands a clear definition of the biologically relevant parameters to be predicted. Biological "compatibility" is obviously too broad because compatibility in different end use contexts depends upon qualitatively different biological parameters. The Materials Science Toxicology Laboratories (MSTL) primary acute toxicity screen has been in use for a number of years for the evaluation of diverse biomaterials and biomedical devices. The screening program has been described by Autian [*1,2*],[2] and earlier aspects of the development of the screen have been reported by Dillingham et al [*3*], Autian and Dillingham [*4*], and Turner et al [*5*]. This effort will be devoted to a discussion of the rationale of the screen, several tests of in vitro/in vivo relationships, and interlaboratory performance of the screening tests.

Rationale

The rationale for the primary screen evolved from the early recognition of the fact that toxic liability of biomaterials is closely related to the presence of leachable toxic substances in the materials. It was concluded, and the conclusion is still valid today, that a screen was needed for the detection of leachable substances. Chemical analysis of extracts of biomaterials was an obvious approach to the problem of detection; however, because of the complexity of extracts, the time and cost involved in analysis, and the abiding question of the relevance of the findings in terms of potential toxic liability, chemical analysis was not pursued as a primary screening procedure. Since a desirable feature of a primary screen is ease and rapidity of execution, and since many biomaterials in use in biomedical applications were known to have an acute toxic liability, acute toxicity was selected as the basis for assessment. In vitro and in vivo tests were devised for testing of materials per se and extracts of the materials; the latter providing confirmation of the leachability of the toxic substance. Although in vitro tests were recognized to be generally more sensitive than in vivo tests, the relevance of in vitro response to in vivo response was not known, a situation largely unchanged

[2] The italic numbers in brackets refer to the list of references appended to this paper.

today. The screen, as a consequence, was comprised of three qualitatively different in vitro tests and three qualitatively different in vivo tests. Two of the tests in each set provided assessment of acute response to extracts.

Since leachable substances of different biomaterials are both quantitatively and qualitatively different, different profiles of acute response across tests are observed. For the comparison of the relative potential toxic liability of different materials, a cumulative toxicity index (CTI) was devised based on the summation of the normalized responses of all the tests. The CTI in conjunction with an assessment of the magnitude of the response in the tests showing positive toxic response was judged to be the best approach to assessment of relative toxic liability.

The Primary Acute Toxicity Screen

The tests comprising the primary screen are presented in Table 1, adapted from Autian [1]. It should be noted that all tests are given equal weight

TABLE 1—*Primary acute toxicity screen, 14 biological responses used to calculate the cumulative toxicity index (CTI)* [1].

Biological Responses	Highest Toxicity Value for Response
1. Tissue culture/agar overlay test on material zone index (0 to 5) + lysis index (0 to 5) × 10	100
2. Tissue culture/agar overlay test on saline extract zone index (0 to 5) + lysis index (0 to 5) × 10	100
3. Tissue culture/agar overlay test on PEG 400 extract zone index (0 to 5) + lysis index (0 to 5) × 10	100
4. Tissue culture/agar overlay test on cottonseed oil extract zone index (0 to 5) + lysis index (0 to 5) × 10	100
5. Intramuscular rabbit implant test-gross observation response (0 to 5) × 25	125
6. Intramuscular rabbit implant test—histopathologic evaluation-response (0 to 4) × 25	100
7. Hemolysis test on material-response (0 to 100) × 1	100
8. Intracutaneous test in rabbits on saline extract-response (0 to 5) × 25	125
9. Intracutaneous test in rabbits on PEG 400 extract-response (0 to 5) × 25	125
10. Intracutaneous test in rabbits on cottonseed oil extract-response (0 to 5) × 25	125
11. Systemic toxicity test in mice on saline extract-response (0 to 5) × 20	100
12. Systemic toxicity test in mice on PEG 400 extract-response (0 to 5) × 20	100
13. Systemic toxicity test in mice on cottonseed oil extract-response (0 to 5) × 20	100
14. Inhibition of cell growth test on aqueous extract-response (based upon nine sample weights totaling 100 points) × 1[a]	100
Total	1500

[a] NOTE—Add 20 for the first point and 10 for each additional point or 29% of higher inhibition.

(maximum of 100 CTI points per test), except for the intramuscular implantation test and the intradermal irritation test (intracutaneous test) where the maximum points per test are 125 CTI points, to give appropriate weight for positive responses that exceed the positive control, a situation encountered with highly toxic materials. The highest CTI value reported to date was 1045 and the lowest 30. Approximately two thirds of all materials have had CTI values of 400 or lower, and approximately one third have had CTI values of 100 or lower.

The primary screen constitutes a tier of relatively simple in vitro and in vivo tests for the presence of leachable toxic substances in biomaterials. Each test provides an independent estimate of the relative toxicity of the material or extract of the material. The four extraction media used constitute a series decreasing in polarity, which assures the extraction of components of widely differing solubility properties. The leachability-toxicity profile as reflected in the magnitude of toxic response across the several tests is characteristic of the particular material tested. Strictly as a matter of judgment based on experience, a material is considered to pose a low potential toxic liability if the CTI value for a material is 100 or lower with no test showing higher than a minimum positive response. It should be noted that approximately four to 6 minimum positive responses out of the 18 parameters scored will produce a CTI of 100, depending upon the distribution of responses across the in vivo and in vitro tests. Seldom are all positive responses limited to either in vitro or in vivo tests.

Characteristics of the Individual Tests

Hemolysis Test

The hemolysis test employs rabbit blood and is designed to determine the hemolytic activity of materials during a 90 min incubation (37°C) of the material in saline with rabbit blood present for the last 60 min. The percent hemolysis is expressed as the percent of the positive control (100% hemolysis) corrected for the negative control (hemolysis in normal saline). The hemolytic activity is assayed spectrophotometrically (545 nm) employing the supernatants obtained after centrifugation (750 times gravity, 5 min), threefold replication throughout. The hemolytic activity of "immediately soluble, leachable" components and the hemolytic activity of the physical surface of the material both contribute to the total hemolytic activity observed.

This test provides an estimate of immediate membrane activity of the leachable components of a material. The relevance of hemolytic activity for in vitro/in vivo prediction has been demonstrated [*5,6*] using quantitative concentration-response (hemolytic activity) correlations with in vivo LD_{50}. It should be noted that hemolytic activity occurs at approximately one order higher magnitude concentration than tissue-culture response (inhibition of cell growth in liquid test systems), and high univariate correlation *r* of 0.98

has been observed between hemolysis and tissue-culture responses. A positive response in the hemolysis test is highly significant in terms of potential cellular level and in vivo acute response in spite of the fact that the test is very sensitive in the concentration-response range, that rabbit bloods differ in sensitivity and that uncontrolled extraction variables influence the quantitative results. Rapidity and ease of extraction make the test valuable from a practical point of view.

Tissue Culture Agar Overlay Test

The tissue culture agar overlay (TCAO) test is designed to test directly for the presence of leachable, toxic substances in solid materials. The test material is placed in contact with the surface of an agar layer overlaying a monolayer of mouse fibroblast cells that are stained with neutral red vital stain. After 24-h incubation, the presence of leachable toxic substances is manifested by the loss of dye from the cells within the diffusion zone of the soluble substance(s) leaching from the specimen and by lysis of the cells within the zone if the concentration and toxicity of the diffusing substance is high enough. The toxic response is reported in terms of a response index based on the size of the diffusion (decolorized) zone and the percentage of the cells within the zone that is lysed.

Extracts or other liquid specimens (imbibed in cellulose assay disks), powders, fibrous materials, metals, and other materials can be tested. The diversity of test materials that can be assayed makes the test the most widely applicable of the primary screen tests and one of the most sensitive tests for leachable, toxic components of materials.

This test has certain characteristics in common with the hemolysis test in that the zone index is based on the magnitude of the diffusion zone of the leachable materials as manifested by loss of vital (neutral red) dye from the cells. This change in differential permeability of the cells is a membrane mediated response as is the hemolytic response. The test differs, however, in that the material specimen never contacts the cells, that the assay time is longer, and that the response can be attributed to leachable, diffusible substances. A second parameter, the percent lysis of cells within the decolorized zone (lysis index), is also scored. This estimate is more subjective than the zone index, although there is significant correlation between the two. In spite of the subjective aspects of quantitation, the TCAO test is one of the more sensitive tests for leachable toxic substances.

Inhibition of Cell Growth Test

The inhibition of cell growth (ICG) test is designed to provide a leachability-toxicity profile based on the inhibition of cell growth of a series of aqueous extracts (decreasing weight/volume) of a biomaterial.

Growth of L-cell monolayer cultures is determined in the presence and absence of the extracts after 72-h incubation. The inhibition of cell growth (percent decrease with respect to negative control growth) is reported for each extract and can be expressed as a leachability-toxicity profile by plotting the percent inhibition against the weight of material extracted.

The potential toxicity of different materials can be compared on the basis of the inhibition observed for equivalent quantities of materials extracted. Highly toxic materials show 100% inhibition even at the lowest quantity of material extracted. A nontoxic material will give no significant inhibition even at the highest level of material extracted.

This test, being based on the inhibition of cell growth as opposed to acute cellular response, is qualitatively different from the other in vitro tests. A difference of 29% inhibition was found to be statistically significant for single-point determinations (for example, for extracts) executed by different operators over time. The test is the most sensitive of the primary screen tests with the possible exception of the intradermal irritation test of cottonseed oil extracts, which however is less quantitative. The relevance of inhibition of cell growth to the rationalization or prediction of in vivo LD_{50} has been demonstrated within chemically related classes of chemicals [*6*]. A significant degree of rationalization of in vitro and in vivo response requires consideration of the physical properties of the toxicants. Neither concentration nor physical properties of leachables are known in extract testing (or direct tests on the materials) and the possibility of synergistic response among leachables is always present. However, a positive response in a screening test provides an unambiguous confirmation of the presence of a leachable substance.

Intradermal Irritation Test

One of the most sensitive of the primary screen tests is the intradermal irritation test. Rabbits are prepared for the test by close clipping of hair from the region of the back. Lightly anesthesized, the rabbits are injected intradermally with 200 μL of a material extract at five sites on the left side about 25 mm (1 in.) from and parallel to the spinal column. The injections begin slightly posterior to the shoulder and extend posteriorly. The negative control (extraction medium) is injected similarly at three sites on the right side beginning slightly posterior to the shoulder, spaced 25 mm (1 in.) apart. The positive control (20% ethanol) is injected at three sites extending posteriorly from the negative control sites with slightly wider spacing. The irritation response is scored at 1, 24, 48, and 72 h based on a comparison of the magnitude of response of control and treatment sites. Ischemia, edema, erythema, and the diameter of the response zone are important parameters in the scoring system. The average maximum response for the treatment sites is reported.

The intradermal irritation response to cottonseed oil extracts is most frequently positive, followed by polyethylene glycol-400 (PEG-400) extracts and saline extracts. Definite positives are highly reproducible, although quantitation of the response is poor as judged by blind studies with known irritants. Although the saline extract is infrequently positive, when it is positive, it is correlated with responses in other tests. The intradermal irritation test is one of the most sensitive in the screen.

Systemic Toxicity Test

The systemic toxicity test is based on a seven-day observation of lethality of standardized extracts of materials injected intraperitoneally (cottonseed oil and PEG-400 extracts) or intravenous (saline extracts) in mice. The test is the least responsive of the primary screen tests; however, a sufficient number of cases have been observed where a positive response (lethality) was present where there was an absence of a marked positive response in the other primary screen tests, which justifies its inclusion in the screen. A high-toxicity material producing a positive systemic toxicity response normally produces a broad positive response across the screening tests. Few materials that are candidates for biomedical applications show such a high level of toxicity, although a few such materials have been detected.

Intramuscular Implantation Test

Rabbits are prepared for the test by close clipping of hair and cleaning of the dorsal surface. A positive control, negative control, and three specimens ("rods" approximately 1 cm long and <1 mm in diameter) are implanted in the paravertebral muscle parallel to the spinal column. A second set of specimens are similarly implanted in the opposite paravertebral muscle. A specimen is implanted by placing it inside a 15 gage or, if necessary, 13 gage hypodermic needle, inserting a stylus behind the specimen, and inserting the hypodermic needle into the central region of the paravertebral muscle of the rabbit. The stylus is held in place, and the hypodermic needle is withdrawn to deposit the specimen in the muscle wound made by inserting the hypodermic needle. Then the stylus is withdrawn. The skin is slipped over slightly, so the puncture site in the skin is not aligned with the implant site in the muscle, and the puncture site in the skin is painted with tincture of Merthiolate.®

Seven days after implantation, the rabbits are sacrificed and the paravertebral muscles exposed (or removed if preferred). Each implant site (controls and specimens) is located. A cut is made through the implant at right angles to the long dimension, to expose a cross section (of the reaction site and adjacent tissue), and the tissue reaction around the implant is scored grossly. An adequate, but not excessive, block of tissue is obtained that contains the implant (reaction) site plus at least 2 mm of surrounding normal tissue. (A block of tissue 1 by 1 cm in cross section, with the implant near its center, is generally adequate.) The blocks of tissue are preserved in formalin and submitted to histopathologic evaluation of the implant sites. Sections cut by 90° to the long axis of the implanted rods are used for the evaluation. Thirteen histopathologic parameters are scored, and a histopathologic rating is assigned. The symmetry and intensity of the tissue reaction around the implant are important considerations in the evaluation, symmetry of response being associated with the diffusion of leachable toxic substances as opposed to physical trauma.

The implantation test is one of the most relevant primary screen tests for the evaluation of candidate biomaterials. Its principal limitations are the time required overall (approximately two weeks) for its execution and the need for technical expertise in the histopathologic evaluation.

Materials Extracts

As a general principle, the absence of leachable substances is desirable for candidate materials entering biomedical applications. Few, if any, of such materials are completely free of leachables. The main objective of the primary screen is to detect the presence of biologically active (toxic) leachables and, an important but secondary objective, to assess the magnitude of the potential toxic liability of the leachables. The screen, as it is structured, functions well as a system for detection. However, extraction conditions can influence the magnitude of observed response, and the relevance of the response to specific end-use conditions is related to extraction conditions. Since any arbitrary set of extraction conditions can mimic end-use conditions in only a limited way, the primary screen was designed to assess the broad potential leachability of a material using accelerated extractions with extraction media of differing polarity. Differential responses to those extracts provide an insight into the lipophilicity of the leaching substances, a property broadly related to toxicity. Although temperature can be expected to affect the rate of extraction, comparisons of the toxicity of aqueous extracts obtained at 37°C (five-day extraction) and 121°C (1-h extraction) in the inhibition of cell growth test showed no significant differences for two materials of moderately high toxicity. This suggests that the observed leachability-toxicity is related to solubility limits of the leachable substances.

Since biological activity is related to the concentration of a toxicant in the aqueous phase, the relevance of aqueous extracts for the assessment of leachability-toxicity of biomaterials is apparent. Overall, the four extraction media employed in the primary screen (water, saline, PEG-400, and cottonseed oil) provide a relatively exhaustive profile of the leachability of a material and probably reflect more accurately than short-term low-temperature extracts the total potential leachability of a material.

Although accelerated extraction conditions are not "physiological," observed toxicity of extracts as assessed in the primary screen is seldom present in the absence of observed toxicity in tests directly on the materials. If tests directly on the material are negative and extract tests are positive, there is always the possibility that degradation of the material contributes toxic leachables. In such cases, which are relatively few in number, 37°C, five-day extraction is recommended to confirm leachability at "physiological" temperature. The practical advantage of accelerated extraction is that it is rapid, relatively exhaustive, and provides sterile extracts immediately available for all tests. Sterility is a requirement for the inhibition of cell growth test and is desirable for all tests.

Correlations of In Vitro and In Vivo Response

General Considerations

In considering the predictive relationships between in vitro and in vivo response, it is worthwhile to consider the state of the art in that area. Direct correlations of in vitro and in vivo response are generally low and, historically, delayed the development of quantitative structure-activity correlation [*7*]. As an example, the direct correlation of inhibition of cell growth ID_{50} with in vivo acute toxicity LD_{50} for acrylate and methacrylate esters was $r = 0.3$ [*8*]. It was recognition of the importance of physical properties of compounds to manifest toxicity or other biological response that led to significant rationalization of biological response in terms of physical properties and, indirectly, the rationalization of in vitro and in vivo response. For the acrylate and methacrylate esters, for example, correlation with physical parameters was $R = 0.97$ (in vitro) and $R = 0.90$ (in vivo) where R is the correlation coefficient of multiple regression. Such correlations depend upon quantitative in vitro and in vivo data and a knowledge of the physical properties of the known pure compounds investigated.

By contrast, biomaterials testing involves tests on complex materials or extracts of those materials. The composition of the material, extracts, and physical properties of the leachable substances in the extracts are not known. The only avenue for the validation of the predictive relationships between in vitro and in vivo response is direct correlation of in vitro and in vivo response where expectation of high correlations is low. The only valid context for the validation of the predictive value of in vitro or any nonclinical test is a coherent, parallel clinical data base. It is the absence of such clinical or "end-use" data bases that constrains relevant validation of the predictive value of in vitro tests. Recognizing the unavailability of "end-use" data bases, the primary screen was structured to provide a tier of in vitro and in vivo tests that have heuristic relevance to potential higher order response and to provide a coherent parallel data base within the screen for the objective examination of in vitro/in vivo correlations.

Correlation Analysis

At the outset, the qualitative nature of the primary screen data base and its semiquantitative characteristics must be recognized. Direct correlations of biological response across the screen confounds chemical composition of the test materials (diverse materials having been tested, including materials that were rendered highly toxic through the introduction of toxic additives) and the qualitative differences among extracts (for example, polar and nonpolar extract tests tend to be negatively correlated). In addition, with the exception of the hemolysis test, quantitation is based on rating scales (Table 1), although the rating scale for the inhibition of cell growth test is based on a continuous variable, percent inhibition. The distribution of ratings for a test across materials is influenced strongly by the distribution of the toxic potential of the materials, essentially uncontrolled in this study except for the deliberate extension of the toxicity range through introduction of additives. Formal analysis of variance indicates a significant deviation from normality of distribution for most tests, depending upon the data base examined. Therefore, the correlation coefficients and statements of statistical confidence obtained cannot be considered rigorous but only suggestive.

The analysis was carried out using stepwise linear regression, correlating individual tests with all other primary screen tests or subsets thereof. The correlations observed and the statistical statements are those generated by the stepwise program and are constrained by the previously mentioned considerations. The data base was comprised of individual test results from 297 primary screens.

Results and Discussion

A summary of the stepwise linear regression of single primary screen tests (Table 1 and 2) on all other primary screen tests is presented in Table 3. Only the first four parameters entering the equations are shown since the addition of further parameters usually did not improve substantially the magnitude of *R*. Parameter coefficients and intercept values are not presented to simplify presentation and because the analysis can not be considered rigorous. The data base consisted of 197 complete primary screens. The regressions were carried out on those screens where the dependent variable was positive, the number of cases being indicated in brackets below the dependent variable. The multiple correlation *R′* obtained using the entire data base (197 complete screens) is also given for comparison with the correlations *R* obtained with the smaller subsets where the tests were positive. The overall correlation *R′* is generally similar to, but lower than, *R*. In vitro tests are clearly correlated with in vivo tests and relatively high correlations appear frequently with the addition of the first or second parameter. The first independent variable entering the equation has the highest univariate correlation with the dependent variable, and other independent variables enter in the order in

TABLE 2—*Primary screen tests—definition of acronyms.*

Acronym	Parameter
	AGAR OVERLAY TISSUE CULTURE TESTS
ZI-M	zone index, material
LI-M	lysis index, material
ZI-C	zone index, cottonseed oil extract
LI-C	lysis index, cottonseed oil extract
ZI-P	zone index, PEG-400 extract
LI-P	lysis index, PEG-400 extract
ZI-S	zone index, saline extract
LI-S	lysis index, saline extract
ICG	inhibition of cell growth
	HEMOLYSIS TEST
H	hemolysis, %
	INTRADERMAL IRRITATION TESTS
ID-C	cottonseed oil extract
ID-P	PEG-400 extract
ID-S	saline extract
	SYSTEMIC TOXICITY TESTS
S-C	cottonseed oil extract
S-P	PEG-400 extract
S-S	saline extract
	INTRAMUSCULAR IMPLANT TESTS
IM	gross pathology rating
HR	histopathologic rating
	ALL PRIMARY SCREEN TESTS
CTI	cumulative toxicity index
	HISTOPATHOLOGIC CRITERIA
DN	degree of necrosis
CN	coagulative necrosis
FD	floccular degeneration
INF	inflammation
PMN	polymorphonuclear leukocytes
LYM	lymphocytes
PC	plasma cells
MAC	macrophages
FEP	fibroendothelial proliferation
F	fibrosis
GC	giant cells
FB	foreign body debris
RS	relative size of involved area
NF	sum of DN, CN, and FD
C	sum of INF, PMN, LYM, PC, and MAC
FG	sum of FEP, F, and GC

TABLE 3—*Regression of single tests giving positive response on all other primary screen tests*[a]

Test	Step 1	Step 2	Step 3	Step 4	Coefficient
ZI-M =	S-P	+ H	− ID-S	+ LI-S	R^{d} = 0.871 (6)[e]
[37][b]	0.770[c]	0.804	0.820	0.841	R' = 0.781
LI-M =	LI-S	− ID-C	+ ZI-M	− ZI-S	R = 0.854 (8)
[57]	0.612	0.698	0.787	0.808	R' = 0.905
ZI-C =	−ZI-M	+ ZI-S	+ LI-P	+ ID-C	R = 1.00 (8)
[19]	0.512	0.788	0.992	0.994	R' = 0.897
LI-C =	LI-M	+ ZI-C	− ID-C	+ S-C	R = 0.941 (10)
[38]	0.547	0.727	0.808	0.847	R = 0.912
ZI-P =	CTI	− ID-S	+ LI-P	− LI-M	R = *[f]
[23]	0.779	0.810	0.827	0.861	R'^{g} = 0.906
LI-P =	LI-M	+ CTI	+ H	− ID-C	R = 0.842 (6)
[44]	0.627	0.750	0.804	0.818	R = 0.914
ZI-S =	LI-S	+ S-P	+ ID-C	+ ZI-P	R = *[f]
[6]	0.481	0.575	0.625	0.641	R' = 0.685
LI-S =	−ID-C	− H	+ ZI-S	+ LI-P	R = 0.836 (5)
[41]	0.530	0.689	0.765	0.827	R' = 0.826
ICG =	IM	− ID-C	− ID-S	− HR	R = 0.917 (13)
[80]	0.703	0.725	0.762	0.774	R = 0.922
H =	−ID-P	+ CTI	− S-C	− LI-C	R = 0.974 (10)
[43]	0.572	0.685	0.834	0.862	R' = 0.862
ID-C =	ZI-P	+ CTI	− S-C	− H	R = 0.966 (12)
[95]	0.875	0.904	0.919	0.929	R' = 0.953
ID-P =	CTI	− H	− ICG	− ZI-C	R = 0.873 (9)
[48]	0.414	0.733	0.764	0.815	R' = 0.941
ID-S =	H				R = 1.00 (1)
[3]	1.00				R' = 0.818
S-C =	LI-P	− LI-S	+ ZI-S	+ LI-M	R = 0.999 (9)
[24]	0.916	0.965	0.976	0.980	R' = 0.967
S-P =	−ICG	+ ZI-S	− ZI-M	+ LI-M	R = 1.00 (4)
[6]	0.745	0.917	0.996	1.00	R' = 0.678
S-S =	−ZI-M				R = 1.00 (1)
[2]	1.00				R' = 0.000 (0)
IM =	CTI	− H	+ LI-P	− ID-P	R = 0.977 (5)
[17]	0.910	0.957	0.966	0.974	R' = 0.917
HR =	IM	+ ICG	+ ZI-S		R = 0.966 (3)
[44]	0.950	0.963	0.966		R' = 0.931
ICG =	IM	− ID-C	− ID-S	− HR	R = 0.917 (13)
[80]	0.703	0.725	0.762	0.774	R' = 0.922
CTI =	S-C	+ ID-C	+ ICG	+ LI-M	R = 0.977 (16)
[197]	0.860	0.934	0.960	0.969	R' = 0.997

[a] See Table 2 for definition of terms.
[b] [ ] contain the number of materials giving positive response.
[c] Numbers below acronyms equal R at each step of multiple correlation through the fourth step.
[d] R = multiple regression coefficient for complete equation.
[e] ( ) = number of test parameters entering the equation.
[f] All cases showing significant positive response (ZI $\geq$ 2) were singular (ZI = 2). The regression equation for all cases is presented.
[g] R' = multiple regression coefficient based on 197 materials.

qualitatively different from the more nonpolar extracts. Note that ICG is correlated very broadly with in vivo response, with only three of the 13 parameters entering the equation being in vitro tests, and the first 9 rank order parameters being in vivo tests. It also appears as the second rank order parameter in the equation for histopathologic rating (HR). This suggests particular relevance of the ICG for prediction of in vivo response. Its relatively high frequency of positive response (41% of the 197 screens) is an indication of its sensitivity.

The protocol for the histopathologic evaluation of the intramuscular implantation tests is based on the assessment of 13 histopathologic parameters (0 to 3 rating scales with the ratings for cellular infiltrates being based on objective cell counts). The HR (0 to 4 rating scale) is based on those data with the greatest weight being given to necrosis and inflammation in the assignment of the rating. The histopathologic data base therefore provides an independent, parallel, higher order in vivo context for the assessment of the individual primary screen tests. Stepwise linear regression of individual tests on histopathologic parameters was carried out using the 13 parameters and three derived values (NF, C, and FG; sums of selected parameters) as independent variables (Table 2). Although, as might be expected, the values of R and R' for those regressions (not presented) are significantly lower than for the intertest regressions of primary screen tests (Table 3), it is informative to consider the dominant parameters entering the regressions and the magnitude of the correlations. Degree of necrosis (DN) and coagulative necrosis (CN) appear first rank order in 14 of the 18 regressions (R ranging from 0.236 to 0.970) with significantly higher correlations with in vivo tests than in vitro tests. The low frequency of positive responses for saline extract tests are again apparent, but their selective high correlation with independent variables (in this case, histopathologic parameters) is apparent as observed in the regressions with primary screen tests (Table 3). The HR shows the expected high correlations ($R = 0.936$), being derived from the assessment of histopathologic parameters. It should be noted that ZI-S (positive response) showed one of the highest correlations with histopathologic parameters ($R = 0.854$). It is also interesting that fibroendothelial proliferation (FEP), which is associated with recovery from toxic tissue response, is the first or second rank order in the regressions of ZI-P, ZI-S, ICG, H (hemolysis test), and IM (intramuscular implantation test, gross rating); compare with rank order correlations for these tests and FEP, Tables 4 to 7. The balance of the histopathologic parameters entering the regressions are cellular infiltrates associated with inflammation, although inflammation (INF) per se appears infrequently, the latter being difficult to assess.

To examine the broad correlation between in vitro and in vivo tests and the correlation between histopathologic parameters and those tests, the individual tests and histopathologic parameters were ranked on the basis of the rank order and frequency of entry into regression equations generated using

TABLE 4—*Correlation of single in vitro tests with all in vivo tests.*

Rank Order	Average Rank Order	Adjusted Average Rank Order[a]
1	ICG	H
2	ZI-M	ICG
3	H	ZI-M
4	LI-P	ZI-S
5	LI-C, ZI-S	LI-P
6	LI-M	LI-M
7	LI-S, ZI-P	LI-C
8	ZI-C	ZI-P
9	...	LI-S
10	...	ZI-C

[a] Adjusted for frequency of improvement of multiple correlation.

the entire data base. Tables 4 to 7 give the rank order for individual tests or histopathologic parameters based on average rank order of entry and based on the average rank order of entry adjusted for frequency of entry (average rank order multiplied by the number of regressions under consideration divided by the number of regressions in which the parameter appeared). The rank order reflects the relative degree of association of a test parameter with in vitro or in vivo tests without respect to the absolute magnitude of the correlation, although correlation decreases as rank order increases; see the R' values for the parameters (Tables 3 and 8) for the correlations with combined in vitro and in vivo tests. Regression equations for the comparisons of Tables 4 to 7 are not presented.

In vitro tests most closely associated with in vivo tests, (Table 4) are H, ICG, ZI-M (TCAO, zone index, material test sample), and ZI-S or LI-P (TCAO, lysis index, PEG-400 extract) based on the first four rank order associations. The appearance of systemic toxicity tests on cottonseed oil (S-C) and PEG-400 (S-P) extracts in the first or second rank order association with in vitro tests (Table 5) reflects the higher order toxicity of nonpolar extracts and the high correlation between in vitro response and systemic response for high toxicity materials. ID-C and ID-P (intradermal irritation tests, cottonseed oil, and PEG-400 extracts, respectively) have the highest correlation

TABLE 5—*Correlation of single in vivo tests with all in vitro tests.*

Rank Order	Average Rank Order	Adjusted Average Rank Order[a]
1	ID-P, S-C	ID-C
2	ID-C	S-P
3	IM	ID-P, S-C
4	S-P	ID-S
5	ID-S	IM
6	HR	HR
7	S-S	S-S

[a] Adjusted for frequency of improvement of multiple correlation.

TABLE 6—*Correlation of single histopathology parameters with all in vivo tests giving positive response.*

Rank Order	Average Rank Order	Adjusted Average Rank Order[a]
1	DN	DN
2	CN, MAC, GC	CN
3	PMN	MAC
4	PC, RS, $\sum$ (FEP, F, GC)	PMN
5	FEP	RS
6	$\sum$ (CN, DN, FD), F, INF	FEP
7	FD, LYM	F, GC
8	...	PC, $\sum$ (FEP, F, GC)
9	...	$\sum$ (CN, DN, FD), INF
10	...	FD, LYM

[a] Adjusted for frequency of improvement of multiple correlation.

with in vitro response. Necrosis (DN and CN) is first rank order with respect to both in vitro and in vivo tests (Tables 6 and 7). Inflammatory cells follow necrosis and there appears to be an inverse differential association of fibroendothelial proliferation (FEP) and macrophage (MAC) with in vitro and in vivo tests. The correlations and rank order associations of tests and histopathologic parameters are in good agreement with observations of test performance in this laboratory. It must be stressed that the correlations presented are based on a finite data base comprised of diverse materials. The magnitude of correlations and rank orders can be expected to change as the data base expands; however, the dominant correlations and rank order associations are not expected to change significantly.

Interlaboratory Performance of Primary Screen Tests

Design of the Performance Test

Six laboratories, including MSTL, were selected for participation on a voluntary basis in the performance test. Representatives of each laboratory at-

TABLE 7—*Correlation of single histopathology parameters with all in vitro tests giving positive response.*

Rank Order	Average Rank Order	Adjusted Average Rank Order[a]
1	CN	CN
2	PMN	FEP
3	FEP	LYM
4	LYM, PC, F, GC	PC, RS, $\sum$ (FEP, F, GC)
5	FD, DN	INF
6	RS, $\sum$ (FEP, F, GC)	F, GC
7	INF	DN, FD
8	MAC	PMN
9		MAC

[a] Adjusted for frequency of improvement of multiple correlation.

TABLE 8—*Regression of single primary screen tests giving positive response on histopathologic parameters.*[a]

ZI-M	= −LYM	− RS	− GC		R^b = 0.379 (3)[c]
[84][d]	0.308	0.353	0.379		R' = 0.507
LI-M	= CN	− GC	+ PC	+ RS	R = 0.405 (5)
[117]	0.236	0.329	0.368	0.385	R' = 0.525
ZI-C	= −GC	+ PMN	− CN	− FG	R = 0.567 (4)
[36]	0.323	0.431	0.525	0.567	R' = 0.530
LI-C	= CN	+ PC	+ FG	− LYM	R = 0.461 (4)
[92]	0.266	0.370	0.422	0.461	R' = 0.459
ZI-P	= CN	− FEP	− GC	+ INF	R = 0.495 (4)
[52]	0.332	0.416	0.459	0.495	R' = 0.695
LI-P	= CN	+ F	+ INF	− FD	R = 0.552 (4)
[113]	0.493	0.510	0.531	0.552	R' = 0.387
ZI-S	= −FEP	+ LYM	+ F	− RS	R = 0.854 (6)
[21]	0.320	0.544	0.702	0.759	R' = 0.387
LI-S	= CN	− FD	+ FG	+ INF	R = 0.433 (4)
[91]	0.293	0.390	0.416	0.433	R' = 0.478
ICG	= DN	− FEP	+ FD	− CN	R = ...[f]
[80]	0.534	0.612	0.645	0.670	R' = 0.678
H	= −FEP	− LYM			R = 0.514 (2)
[100]	0.475	0.514			R' = 0.397
ID-C	= DN	+ PMN	− FG		R = 0.817 (3)
[115]	0.771	0.799	0.817		R' = 0.635
ID-P	= CN	− MAC	+ DN	− NF	R = 0.662 (5)
[68]	0.556	0.598	0.625	0.645	R' = 0.584
ID-S	= −DN	+ GC			R = 1.00 (2)
[4]	0.927	1.00			R' = 0.303
S-C	= DN	+ RS	+ PC	− INF	R = 0.873 (6)
[42]	0.733	0.814	0.836	0.855	R' = 0.786
S-P	= CN				R = 0.364 (1)
[17]	0.364				R' = 0.437
S-S	= CN	− DN			R = 1.00 (2)
[3]	0.970	1.00			R' = 0.208
IM	= DN	+ FEP	+ F		R = 0.865 (3)
[36]	0.832	0.857	0.865		R' = 0.860
HR	= DN	− MAC	+ PMN	− RS	R = 0.963 (6)
[77]	0.947	0.951	0.957	0.960	R' = 0.947

[a] See Table 2 for definition of terms.
[b] Numbers below acronyms equal R at each step of multiple correlation through the fourth step; parameter coefficients and intercept values not shown.
[c] ( ) = number of test parameters entering the equation.
[d] [ ] contain the number of materials giving positive responses.
[e] R' = multiple regression coefficient based on 277 materials.
[f] Not determined.

tended an orientation conference at MSTL before initiation of the performance tests. The detailed protocols, standardized reporting forms, and test materials were provided by MSTL who was responsible for evaluation of the performance tests.

The primary screen tests were executed by each laboratory on four materials selected by MSTL to provide a moderate range of toxicity (CTI values ranging from approximately 100 to 400). The primary screens were executed "blind" and in random order with fivefold replication to provide a data base for an analysis of variance of the test data for individual tests and laboratories. Recognizing the limitations of parametric analysis, previously discussed, both parametric and nonparametric analysis were employed in the evaluation.

Results and Discussion

The results of the statistical analysis of the performance data are presented in Table 9. It should be noted that one laboratory (Laboratory II, see Figs. 1 and 2) completed only two primary screen runs on each material, and data from that laboratory was therefore excluded from the analysis. Also, the data for the intradermal irritation test for the ID-C for two laboratories (Laboratories V and VI) was excluded in the evaluation of that test because of observed toxicity of the cottonseed oil extraction medium (negative control) used. Mean values based on the performance of all other laboratories were used in the calculation of CTI for those laboratories. This did not affect the

TABLE 9—*Statistical analysis—primary screen tests.*

Test	Main Effects							
	Source of Variation[a]					Interaction		
	Labs (A)		Materials (B)		Time (C)	AB	AC	BC
	P	NP	P	NP	P	P	P	P
ZI-M	*[b]	—[c]	*	*	—	*	—	—
LI-M	*	—	*	*	—	*	—	—
ZI-C	*	*	*	+[d]	*	—	*	—
LI-C	*	*	*	+	+	*	—	—
ZI-P	*	*	*	+	+	—	—	—
LI-P	*	—	*	+	—	*	—	—
ZI-S	*	—	*	+	—	*	—	—
LI-S	*	—	*	+	—	+	—	—
ICG	*	+	*	+	—	+	—	—
H	*	+	*	+	—	+	—	—
ID-C	*	—	*	—	—	*	—	+
ID-P	*	+	*	*	—	—	—	—
ID-S	+	—	—	—	—	—	*	+
S-C	...	...	...	...	...	...	...	...
S-P	...	...	...	...	...	...	...	...
S-S	...	...	...	...	...	...	...	...
MI	*	—	*	—	—	+	*	—
HR	*	*	*	—	+	+	—	—
CTI	*	+	*	*	—	*	—	—

[a] A, B, and AB interaction obtained from analysis without time as a main effect; P = parametric analysis and NP = nonparametric analysis.
[b] * = significant, ≥99% confidence.
[c] — = nonsignificant, <95% confidence.
[d] + = significant, ≥95% confidence.

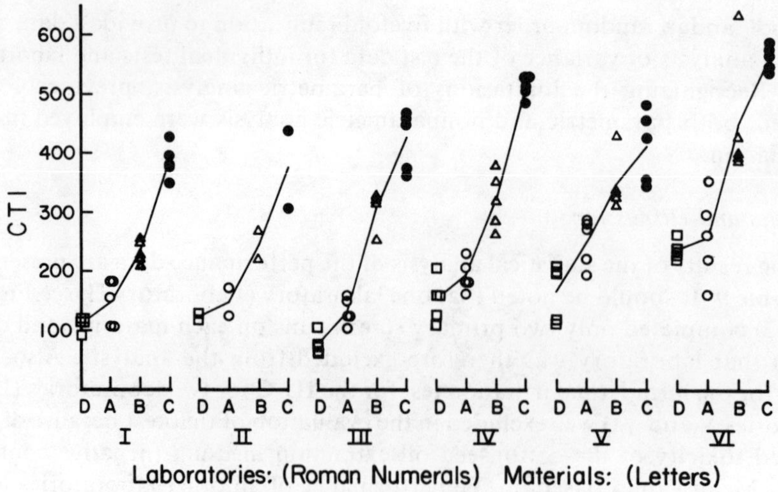

FIG. 1—*CTI values obtained on four test materials in the interlaboratory evaluation of the primary screen. (Laboratory I is MSTL.)*

analysis for other laboratories or other tests, and the procedure did not affect the reported CTI values for laboratories V and VI significantly. Otherwise, all test ratings were used as received without respect to nonstandard control values reported in a low percentage of the tests (agar overlay tissue culture test, 8.6%; intradermal irritation test [all tests not excluded as previously described], 2%; and intramuscular implantation test, 2.4%). The analysis, therefore, was based on maximum expected variation of the tests. The parameter of "time" was the order of testing over time, not real time between tests, which varied substantially among laboratories.

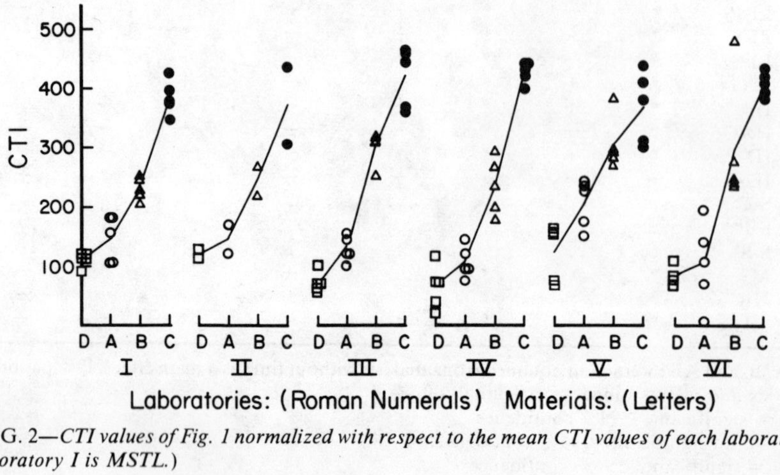

FIG. 2—*CTI values of Fig. 1 normalized with respect to the mean CTI values of each laboratory. (Laboratory I is MSTL.)*

The only test that did not discriminate significantly between materials was intradermal irritation, ID-S, although ID-C, MI, and HR were not significant by nonparametric analysis, a technically more valid but less sensitive analysis. Order of testing ("time") was significant in only three cases ZI-C (TCAO, zone index, cottonseed oil extract), ZI-P, and HR. Only HR showed a consistent trend with respect to time for all four materials, a slight decline that may be related to increasing experience with the histopathologic rating system not used previously by any of the laboratories except MSTL. Differences between laboratories were significant for all tests, although nonparametric analysis discriminated significant differences in only 7 out of 19 tests. There was significant interaction between laboratories and materials for most of the tests but few interactions between materials or laboratories and "time." Similar results were obtained from an analysis of variance of histopathologic ratings, except that only 8 out of 17 parameters significantly discriminated between materials. It is significant to note that those parameters included DN, CN, INF, MAC, GC, and two derived parameters (sums). This objectively supports the use of necrosis and inflammation in the assignment of HR.

The overall performance of the primary screen in the six laboratories is shown in Fig. 1. Figure 2 shows the CTI values normalized with respect to the mean CTI value for each laboratory. All six laboratories consistently discriminated the two higher toxicity materials (B and C) from the two lower toxicity materials (D and A), although absolute values differed significantly between laboratories. Based on the observed within laboratory variance of CTI values, the least significant difference (95% confidence) for CTI values within a single laboratory was 123 CTI units.

The performance trial clearly indicated the utility of the primary screen tests for the detection of leachable toxic substances of biomaterials. It also indicated that single laboratories using the primary screen protocols can determine the relative toxicity of materials with a useful degree of discrimination. All laboratories ranked the toxicity of the test materials in the same order. However, determination of laboratory performance with respect to standard reference materials will be required for meaningful comparison of CTI values generated in different laboratories. Although the predictive value of in vitro tests for the prediction of in vivo or "end-use" response cannot be stated quantitatively at the present time, demonstration of significant correlations between in vitro and in vivo tests provides a basis for continuing development of predictive systems.

Acknowledgments

This work was supported in part by the National Heart, Lung, and Blood Institute, Grant Number R01-HL-24040, Contract Number N01-HV-82914, and the Food and Drug Administration, Contract Number 223-73-5231.

References

[1] Autian, J. in *Artificial Organs,* Vol 1, 1977, pp. 53–60.
[2] Autian, J. in *Fundamental Aspects of Biocompatibility,* Vol. II, D. F. Williams, Ed., Chemical Rubber Company Press, Inc., Fla., 1981, pp. 63–86.
[3] Dillingham, E. O., Webb, N., Lawrence, W. H., and Autian, J., *Journal of Biomedical Materials Research,* Vol. 9, 1975, pp. 569–596.
[4] Autian, J. and Dillingham, E., *Medical Instrumentation,* Vol. 7, No. 2, 1973, pp. 125–130.
[5] Turner, J. E., Lawrence, W. H., and Autian, J., *Journal of Biomedical Materials Research,* Vol. 7, 1973, pp. 39–58.
[6] Dillingham, E. O., Mast, R. W., Bass, G. E., and Autian, J., *Journal of Pharmaceutical Sciences,* Vol. 62, No. 1, 1973, pp. 22–30.
[7] Hansch, C. and Leo, A., "*Substituent Constants for Correlation Analysis in Chemistry and Biology,*" Wiley, New York, 1979.
[8] Dillingham, E. O., Schmalz, G., Lawrence, W. H., and Autian, J., "*Acrylate and Methacrylate Esters: In Vitro/In Vivo Toxicity Relationships,*" Program: Eighth Annual Meeting of the Society for Biomaterials and Fourteenth International Biomaterials Symposium, Paper No. 26, 1982, to be published in *Journal of Biomedical Materials Research,* 1983.

Thomas F. Genova,[1] *Stanko Bodnar,*[1] *Maryellen Vaccarella,*[1] *Carol Stafford,*[1] *and Kathleen M. Schady*[1]

Cytotoxicity Testing: Prediction of In Vivo Toxicity from In Vitro Tests

REFERENCE: Genova, T. F., Bodnar, S., Vaccarella, M., Stafford, C., and Schady, K. M., **"Cytotoxicity Testing: Prediction of In Vivo Toxicity from In Vitro Tests,"** *Cell-Culture Test Methods, ASTM STP 810,* S. A. Brown, Ed., American Society for Testing and Materials, 1983, pp. 71–76.

ABSTRACT: Tissue culture laboratories can play key roles in the development and evaluation of products. Three broad categories of materials are evaluated: new product components (polymers, fabrics, and adhesives), candidate packaging materials (papers, plastics, coatings, and inks), and existing products undergoing stability studies. Material cytotoxicity is evaluated with mouse fibroblasts (L-929) using several standardized tests: agar overlay, elution assay, and direct growth assay. The severity of specimen testing is governed by the intended use of the material. Testing severity decreases in order of the following end use categories: implantations, limited patient contact, indirect patient contact, and no patient contact.

Sample cytotoxicity is compared with that of standard, calibrated controls. Controls are formulated such that they evoke minimal responses in United States Pharmacopeia toxicity tests. Each lot of control material is certified via in vivo analysis before being used in cell culture. Control responses in vitro are then used to predict in vivo toxicity. Since the control is known to evoke a minimal in vivo response, its cell-culture response may be assumed to be equivalent to that of minimal in vivo toxicity. It is assumed that any material having a cell-culture response greater than the control will be toxic when tested in vivo. Studies are underway to further refine the correlation between in vivo and in vitro test methods.

KEY WORDS: tissue culture, cytotoxicity, in vivo toxicity

Tissue culture laboratories may play key roles in evaluating the biocompatibility of materials used or being considered for use as product components. Both the manufacturing and research divisions of a company may receive support from these laboratories. For the manufacturing division, investigations may take the form of potential contraindications regarding the use of new vendors or manufacturing process parameters or both. The research division receives support through both a biocompatibility screening program and research programs that develop in vitro models of in vivo reactions.

[1] Section manager, assistant scientist, senior technician, research assistant, and manager of microbiology, respectively, ETHICON, Inc., Somerville, N. J. 08876.

Specimens being evaluated by the laboratory may be classified into one of several broad categories based on the degree of specimen/product development and the intended end use of the material (Table 1). Existing products, which have already passed a complete biocompatibility screen, are tested for cytotoxicity during accelerated aging studies thereby facilitating the detection of potential toxicity resulting from the degradation of product components [*1,2*].[2]

Candidate packaging materials and product components are evaluated more thoroughly to preclude the possibility of adverse product/package or product/patient interactions.

The first level of testing involves those materials that have no patient contact, for example, packaging materials. Specimens in this category are evaluated via agar overlay testing using the response rating system defined by Rosenbluth et al [*3*] and Autian [*4*]. Extracts prepared using United States Pharmacopeia (USP) surface to volume parameters are tested to determine the potential for transfer of toxic moieties from package to product [*5*].

The second level of testing involves materials coming into topical contact with a patient. Items in this category include fabrics, adhesives, instruments, and so forth. Material cytotoxicity is determined via the agar overlay test. Material in this category may be subjected to extraction by the patients' body fluids. For this reason the screening process includes tests on extracts prepared in tissue-culture media incubated at 37°C for 24 h.

The third level of testing involves materials and products that come into

TABLE 1—*Cytotoxicity test program.*

Degree of Patient Contact	Example	Test[a]	Specimens: Material	Specimens: Extract Type[b]
None	packaging materials	AO	+[c]	NS
Topical	fabrics	AO	+	TCM
Topical	adhesives	. . .	. . .	. . .
Topical	instruments	. . .	. . .	. . .
Indirect	transfusion assemblies	AO	+	TCM,CSO
Indirect	syringes	. . .		
Indirect	catheters	P	NT[d]	NaCl
Internal	sutures	AO	+	TCM, CSO, H_2O
Internal	ligating clips	EA	NT	TCM
Internal	. . .	DGA	+	NT
Internal	. . .	P	NT	NaCl

[a] Test: AO = agar overlay; EA = elution assay; DGA = direct growth assay; and P = pyrogenicity test.

[b] Extract: TCM = tissue culture media; CSO = cottonseed oil; H_2O = water; NaCl = 0.85% sodium chloride solution; and NS = nonstandard (for example, packaging fluids).

[c] + = tested.

[d] NT = not tested.

[2] The italic numbers in brackets refer to list of references appended to this paper.

indirect contact with a patient. Examples of this category include devices used to administer parenteral solutions (for example, transfusion assemblies). Since these devices are subjected to a continuous extraction process, any material (for example, plasticizers) extracted from them can have a potentially adverse effect on the patient. The presence of extractable toxic components is determined by preparing extracts in both polar and nonpolar solvents, that is, tissue culture media and cottonseed oil. These extracts and solid specimens of the materials in question are tested using the agar overlay test. Pyrogen tests are also conducted on these devices to preclude the possibility of adverse patient reaction caused by endotoxin contamination.

The fourth level of testing involves materials or devices that are intended for internal usage, for example, sutures, ligating clips, and so forth. All devices in this category are extracted in tissue-culture media containing 10% heat inactivated fetal bovine serum and cottonseed oil. The extracts and devices are then tested via the agar overlay test. Two other assays are also used in the evaluation process: (1) extracts in tissue-culture medium are tested via an elution assay since device extracts may be metabolized by the cells and undergo metabolic activation to become a toxic material [*6*] and (2) a direct growth assay is used to examine the effects of the device in close contact with cells [*7*]. Finally, devices in this category are tested for pyrogenicity to ensure the absence of endotoxin contaminants [*5*].

The underlying assumption in all of these tests is that they are accurate predictors of in vivo reactivity. The predictive interpretation of cell-culture data have often been one sided, that is, materials that are nontoxic in vitro are assumed to be nontoxic in vivo. This is a conservative interpretation and is based upon the sensitive nature of cell-culture systems. The goal of these tests is to aid in the prediction of in vivo toxicity from in vitro tests. ETHICON, Inc. has approached this goal via a series of experiments designed to correlate in vitro test results with those of seven day intramuscular implant tests.

Intramuscular implantation of medical devices is known to elicit a variety of cellular and enzymatic responses [*8*]. We have utilized several variations of the direct growth assay to study these parameters. Viability tests may be done wherein fibroblasts in direct contact with the device are stained with neutral red and examined for viability and cytopathology. The direct growth assay is also used to study acid phosphatase activity in fibroblasts and macrophage cultures grown on specimens [*9*]. Lysosomal enzyme activity in response to intramuscular implants has been demonstrated by others [*9*].

Another approach to the problem of in vivo/in vitro correlation is through the use of chemically and biologically defined controls [*10*]. Materials used as standard controls must meet several key requirements:

(1) They must be readily available or easily formulated from readily available materials

(2) They must be chemically defined

(3) They should elicit a reproducible response in the chosen assay system
(4) They must be stable after storage

(1) Control materials must be readily available. It would be disastrous to develop a standardized procedure and then discover that your standard control was no longer available. (2) The need for a chemically defined material reflects not only the first requirement (availability), but also the requirements of the current Good Laboratory Practices regulations. (3) The need to elicit reproducible responses in controlled assays is critical since this material is to be used as a routine predictor of biological activity. (4) The need for storage stability is dictated both by the requirement for reproducibility and the need to prepare large quantities of material with a minimum time investment.

We found that films of polyurethane meet these requirements. Thermolite 31, an organotin plasticizer, is incorporated into the films to produce a toxic response. Film toxicity was determined in both the agar overlay and USP intramuscular implantation tests. Dose response curves for both assays indicate a good correlation between reactivity and plasticizer concentration (Fig. 1). This correlation r was better in the intramuscular test than in the agar overlay test; r^2 equals 0.985 and 0.939, respectively.

A direct comparison of the two systems, in vivo and in vitro, may be obtained by plotting intramuscular versus agar overlay responses (Fig. 2). Reactivity in the two systems is linearly related; thus, increases in intramuscular toxicity are paralleled by increased toxicity in the agar overlay test

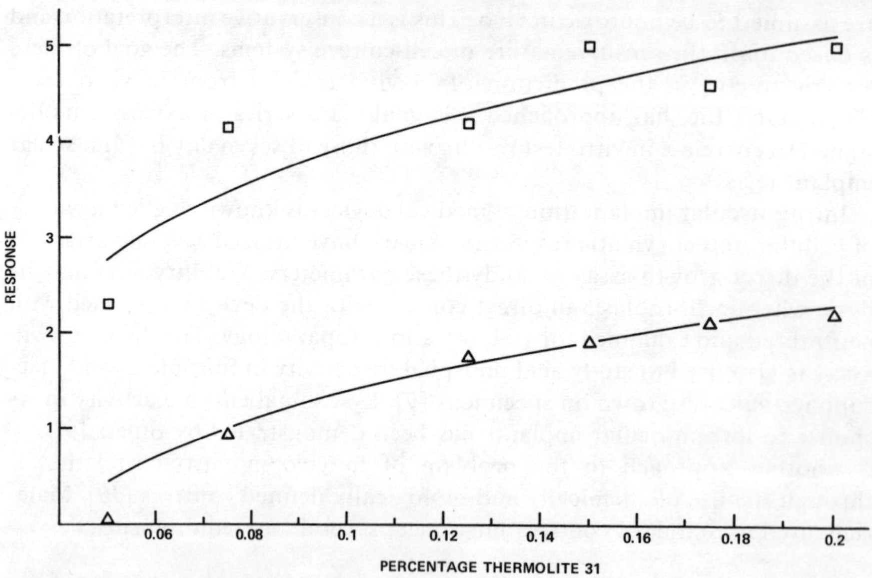

FIG. 1—*Dose response curves for Thermolite 31 in polyurethane.* △ = *intramuscular response and* □ = *tissue culture response.*

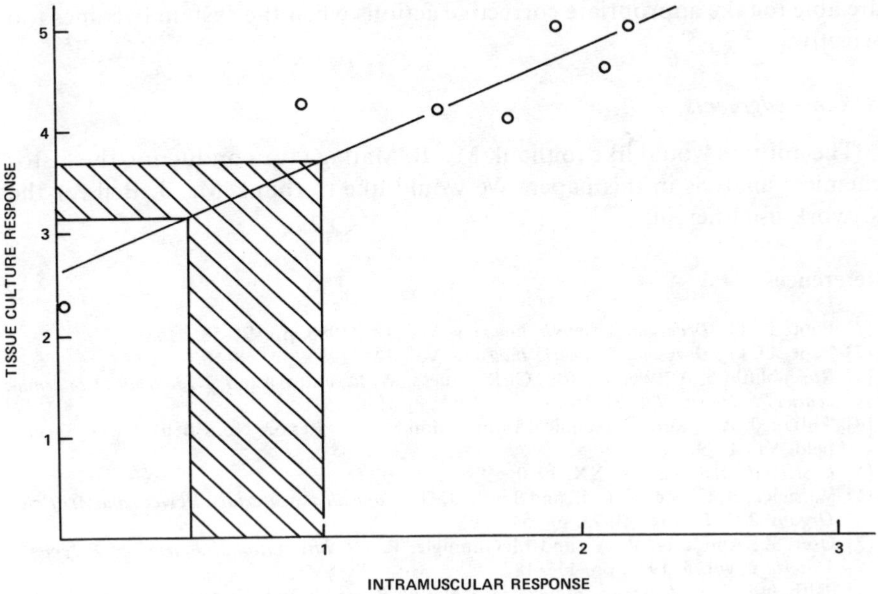

FIG. 2—*Correlation of in vivo versus in vitro responses.*

($r^2 = 0.914$). The crosshatched areas of the graph represent a point of response equivalency between the in vivo and in vitro assays. This point represents the minimum response detectable in intramuscular assays, that is, slightly toxic. In our laboratories, this response (0.5 − 1.0 in vivo) is equivalent to a response of 3 − 4 in vitro. Because of the linear correlation between in vivo and in vitro test methods, any specimen having a response equal to or greater than the minimal animal response equivalent (MARE) in vitro would be expected to be toxic in vivo. Specimens having a response less than the MARE in vitro would not be expected to be toxic in vivo.

In order to avoid running a dose response curve each time the agar overlay test is routinely performed, we have chosen as our positive control a material that reproducibly elicits a slight response in vivo (0.075% Thermolite 31). Since this control is minimally toxic in vivo, its cytotoxic response in vitro is representative of minimal intramuscular toxicity as previously discussed. Thus, by comparing the cytotoxic response of a specimen to that of the control, we are able to judge the specimen's ability to elicit a toxic response in vivo. We believe this model enhances our ability to predict in vivo toxicity from in vitro assays. The use of a standard control eliminates ambiguities arising from day to day variations of in vitro responses since specimen toxicity is judged only on the basis of a comparison with our control response. In addition to precluding our concern over daily response variations, this control allows us to monitor the overall sensitivity of our test system so that we are able to take appropriate corrective actions when the system becomes too sensitive.

Acknowledgments

The authors would like to thank Ms. B. Matlaga for conducting the histochemical analysis in this paper. We would like to thank Mr. J. Bell for the artwork used herein.

References

[*1*] Pope, D. G., *Drug and Cosmetic Industry,* Vol. 127, 1980, pp. 54–58.
[*2*] Pope, D. G., *Drug and Cosmetic Industry,* Vol. 127, 1980, pp. 48–58.
[*3*] Rosenbluth, S. A., Weddington, G. R., Guess, W. L., and Autian, J., *Journal of Pharmaceutical Sciences,* Vol. 54, No. 1, Jan. 1965, pp. 156–159.
[*4*] Autian, J. A., National Technical Information Service, Report No. PB80-141831, Springfield, Va., 1979.
[*5*] *U.S. Pharmacopeia,* Vol. XX, 1980.
[*6*] Wilsnack, R. E., Meyer, F. J., and Smith, J. G., *Biomedical Materials, Devices and Artificial Organs,* Vol. 1, No. 3, 1973, pp. 543–562.
[*7*] Oser, Z., Abodeely, R. A., and McGunnigle, R. G., *International Journal of Polymeric Materials,* Vol. 5, 1977, pp. 177–187.
[*8*] Salthouse, T. N., *Journal of Biomedical Materials Research,* Vol. 10, No. 2, March 1979, pp. 197–229.
[*9*] Salthouse, T. N. and Matlaga, B. F., *Journal of Surgical Research,* Vol. 19, No. 2, Aug. 1975, pp. 127–132.
[*10*] Wade, C. W. R., Hegyeli, A. F., and Kulkarni, R. K., *Journal of Testing and Evaluation,* Vol. 5, No. 5, Sept. 1977, pp. 397–400.

John F. Lontz,[1] *Mary D. Nadijcka,*[1] *and Richard Holmes*[1]

Assessment of Biocompatibility of Orofacial Materials and Devices by Culturing with Human Excised Donor Tissues

REFERENCE: Lontz, J. F., Nadijcka, M. D., and Holmes, R., **"Assessment of Biocompatibility of Orofacial Materials and Devices by Culturing with Human Excised Donor Tissues,"** *Cell-Culture Test Methods, ASTM STP 810,* S. A. Brown, Ed., American Society for Testing and Materials, 1983, pp. 77–87.

ABSTRACT: Three methods for assessing biocompatibility of materials used in orofacial constructs have been studied. These methods are (1) human excised donor (HED) orofacial tissues cultured in medium containing processed human serum, (2) an established cell line cultured in a serum free medium, and (3) an established human cell line cultured in a medium containing processed human serum. Of the three methods, the first was preferred since the cells used were removed recently from their total human physiology and may provide the most realistic evaluation of human-tissue response.

KEY WORDS: cell culture, biocompatibility, human serum, growth factors, toxicity

The biocompatibility tests in this study use a culturing system in which both cells and serum or serum derived additives are species homologous, that is, human. The chemically defined test medium A3 has been formulated specifically for nutrition of human cells [*1*].[2] It is believed that an all human test system should render results that are more likely to reflect the reaction of human physiology to the material composing the orofacial constructs rather than by using the mixed species culturing procedures currently in practice.

Although three optional tests are offered, the one considered potentially to deliver the most humanly relevant data is the human excised donor test (HED). In this proposed option, the cells are temporarily close to their original environment and may carry over substances that provide a more true assessment than the other two test options that use established cell lines. The two optional established cell-line test systems either use human serum or the alpha-protein growth factor [*1*] as a medium additive.

[1] Associate director, research biologist, and biochemist, respectively, Maxillofacial Prosthetics Center, Veterans Administration Medical Center, 1601 Kirkwood Highway, Wilmington, Del. 19805.

[2] The italic numbers in brackets refer to the list of references appended to this paper.

The human serum, used as an additive, is processed from whole blood by centrifugation and dialysis. The alpha growth factor is separated from this processed serum by column chromatography using glass microbeads [*2*]. The test material as well as a toxic and a nontoxic reference material are cultured for not less than 5 but no more than 30 days, depending upon the time required to reach cell confluency in the culture. The reaction of the cells to the materials is assessed by morphology, cell count (growth), or by any appropriate biochemical marker, such as an enzyme.

Materials and Methods

Human Cell Types

Two general types of human cells are used: (1) first passage cells derived from orofacial explants and (2) an established cell line of normal noncarcinogenic origin. The orofacial tissue is obtained from tissues excised from patients during surgery. The tissue is placed in a petri dish on sterile gauze moistened with culture medium. The tissue pieces are then transferred to a small volume of medium. (Volume used depends upon the size of tissue pieces; only enough liquid to prevent drying during the preparation should be used.) Pieces of tissue approximately 1 to 2 mm are cut. The freshly cut tissue pieces are then placed in an inverted flask, along with an appropriate volume of medium and serum, and suspended using a spatula from the top (bottom of the flask). The flask is incubated at 37 ± 0.5°C (98.6 ± 1.7°F) in a humid atmosphere composed of 5 parts carbon dioxide and 95 parts air for a minimum of 2 or a maximum of 4 h. The flask is then carefully turned upright and incubated for 10 to 14 days before being checked for growth. The culture medium is renewed at 48 h intervals. A record of donor age and race as well as type or site of tissue or both is kept. These data will be stored for future possible correlation of test results with these variables.

The established cell line used in this laboratory was the Girardi human heart cell line [*3*]. However, because of the possible contamination of the stock with HeLa cells [*4*], its use has been discontinued. The use of an established cell line has the convenience of rapidly monitoring the quality of the medium, human serum, activity of the growth factor, and operator performance.

Processed Human Serum

Whole blood (60 to 200 mL per pack) within 48 h of the expired shelf life is procured from the Blood Bank of Delaware. Serum is obtained by allowing whole blood to settle at 2 to 4°C (35 to 39°F) in the original blood pack for at least 48 h. The plasma is decanted and centrifuged at 12 000 rpm (Sorval RC-2) for 20 min at 20°C (68°F) to remove any residual blood cells. The plasma is then dialyzed against a balanced salt solution consisting of 84 g NaCl, 4.8 g

KCl, 2.4 g $MgSO_4 \cdot 7H_2O$, 1.68 g $CaCl_2 \cdot 2H_2O$, 1.37 g $Na_2HPO_4 \cdot 7H_2O$, and 0.72 g KH_2PO_4 in 12 L of deionized or distilled water or both at 4°C (39°F). The dialysis is carried out in Spectrapor membrane tubing (Spectrum Medical Industries, Inc.) of 3500 molecular weight exclusion. The method removes small molecular weight toxicants or growth inhibitors producing a serum that can be used as a nutrient supplement for tissue culture.

Cell Growth Factors

Serum-free growth may be sustained by using Holmes alpha-protein growth factor with an established cell line. This growth factor is available from GIBCO Laboratories or may be manufactured in the laboratory according to the published methodology [*2*]. Eventually other growth factors may be used in order to sustain growth of specific cell lines derived from primary tissues in serum-free media [*5*].

Chemically Defined Medium

Holmes A3 medium [*1*] is used for all test purposes. This medium, formulated specifically for use with human cells, has attempted to reintroduce many of the small molecules that have been eliminated by dialysis. For example, in addition to including the water soluble vitamins, fat soluble vitamins are added in the form of a colloidal suspension.

Test Materials

Three general types of test materials are used:

(1) nontoxic reference ("negative" control), material that produces no morphological change in cultured cells;

(2) toxic reference ("positive" control), material that produces a maximum toxic response, under such a condition the culture is almost devoid of cellular growth, most cells of the original inoculum have died; and

(3) test material, in an appropriate form, taken from both stock material and fabricated prostheses. Specimens should be taken from new prostheses before clinical application and from worn prostheses if any complaint arises.

The preferred nontoxic reference used in this test system is Teflon® fluorocarbon (FEP) film available as a 10-mil (0.010-in.) sheeting (E. I. du Pont de Nemours and Co., Inc.). Other polymeric materials can be used as nontoxic references including grades of polystyrene from which a wide range of tissue-culture flasks and devices are made commercially. Two toxic reference materials have been used in our laboratory. Both polyurethane [*6*] and poly(vinylchloride) plastisol [*7*] display maximum toxic responses [*8*]. Since different sources and lots of polymers thought to be maximally toxic may yield conflicting results, a technique for circumventing this is suggested. Our laboratory has manufactured its own toxic reference materials by milling into the nontoxic reference materials substances from the *Toxic Substances* List [*9*]. For example, 0.2% of 4,6 dinitrocresol in polydimethylsiloxane has been used as a toxic reference.

Test Arrangement

The test system involves two principal technical requirements when materials are placed in culture, namely, (1) materials should be retained adjacent to the surface of the confluent layer of cells and (2) materials should be sufficiently transparent to be viewed microscopically during the culture period. For the first of these, the test materials must be immobilized so as not to (1) disturb the growing monolayer or (2) float thereby losing direct contact with the cells, a necessary replication of the in vivo tissue-prosthesis relationship.

High density Teflon (FEP) sheeting (2.2 kg/m^3) and moderately high-density polydimethylsiloxane (1.2 to 1.4 kg/m^3) can reside immobilized on the bottom of the culture flask. However, low or moderate density hydrocarbon polymers (0.91 to 0.94 kg/m^3), certain acrylics (1.17 to 1.20 kg/m^3), and porous test materials tend to float in media of slightly higher density. Therefore, a mechanical arrangement is provided as shown in Fig. 1. A pair of slots is cut in the test material, and a suitably cleaned cover glass (9 by 50 mm, No. 1) is threaded through the slots. A round hole (3 mm in diameter) is cut in the center of the test material in order to record the test photomicrographically. The assembled test material-glass coverslip arrangement is autoclaved before inserting into the culture vessel.

Test Procedure

The sterile test material assembly is inserted into either a Leighton tube (16 by 83 mm) or a petri dish (60 by 15 mm). Cells for test purposes are

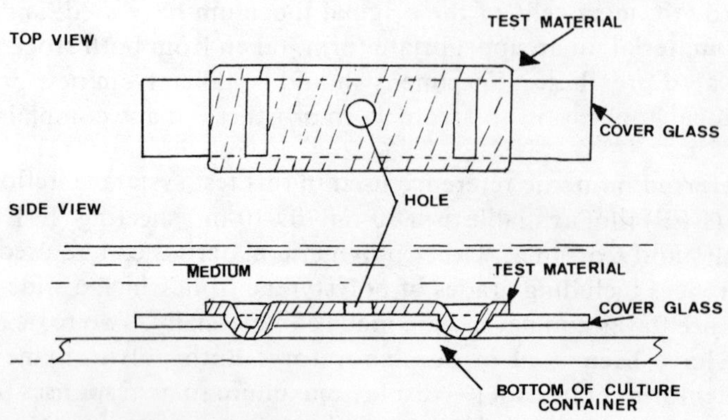

FIG. 1—*This test arrangement is used in the cell culture biocompatibility test system to provide total immersion and immobilization of orofacial materials and prosthetic devices.*

grown in 25-cm^2 flasks until confluency. Cells from established cell lines are removed by scraping with a silicone scraper. Cells derived from primary cultures are trypsinized using a 0.25% trypsin solution. Other enzyme solutions may be used as necessary. Four test vessels are seeded from one confluent flask. The cells are suspended in medium sufficient to furnish 4 mL for each Leighton tube or 5 mL per petri dish. In this manner, the cells are randomly dispersed so that they may grow on the cover glass or directly on the test material. Any air trapped under the test material is expelled by means of a sterile pipette. Cultures are maintained in a humid atmosphere composed of 5 parts carbon dioxide and 95 parts air at 37°C ± 0.5°C (98.6 ± 1.7°F). Medium is changed at 48-h intervals. Time in culture depends upon the type of cell and its frequency of cell division. The test is normally terminated once confluency has been reached. However, a minimum of 5 days (usually met by an established cell line cultured with Holmes alpha growth factor or serum) and a maximum of 30 days (first passage of cells from HED tissues cultured with human serum usually takes 10 to 20 days) are recommended.

Table 1 depicts the design of the test using the minimum number of cell cultures. Test cultures are checked daily for growth and signs of toxicity. At the termination of the test, photomicrographs showing an edge of the test material in contact with cultured cells are taken. Permanent slides may be made.

A modification of a technique used to make permanent slides from embedded tissue is used. Coverslips with test materials are removed from the culture flasks, washed in two changes of a balanced salt solution, and fixed in neutral buffered formalin [*10*] for 30 min. The slides are then run through a series of short stainings and dehydrations: tap water rinse, 1 min; hematoxylin [*10*], 30 min; tap water rinse, 15 min; 50% ethanol, 1 min; 75% ethanol, 1 min; 95% ethanol, 1 min; absolute alcohol, 1 min; absolute alcohol:toluene (1:1), 1 min; and toluene, 1 min. Coverslips are then attached to clean glass slides with Permount.®

For the optional biochemical assessment, cultures are washed with A3 medium containing no additives, scraped from the flask, and stored in small

TABLE 1—*Test design.*

Series	Description	Morphological Assessment	Optional Biochemical Assessment
a[a]	no material	4	4
b	prosthesis specimens	4	4
c	toxic reference	4	4
d[b]	nontoxic reference	4	4

[a] Series a serves to assess such variables as donor source, environment, and operator technique.

[b] Series d may be omitted as a redundant series over Series a in which nontoxic plastic labware is used.

NOTE: numbers may be increased as needed for statistical or other purposes.

volumes at an ultralow temperature, −62°C (−80°F), until such assessment can be completed.

Materials should be considered cytotoxic if there is visible cell lysis or death or both especially in the vicinity of the specimen. A marked reduction in cell number as determined by cell counting or deoxyribonucleic acid (DNA) analysis also serves as an indicator of a toxic response. If specific biochemical analyses (chosen for certain cell types) deviate from normal values, this may be an indication of nonbiocompatibility.

Results and Discussion

Laboratory experience has demonstrated the variability of serum lots in culture [*10*]. At our laboratory, when using fetal-calf serum in some of our earlier test systems, we have noted similar fluctuations in quality. Table 2 displays data derived from some of our earlier tests using fetal-calf serum.

A major alteration in the test system was made when fetal-calf serum was replaced by human serum. Human serum displayed the same type of variability from lot to lot as the fetal-calf serum. Although not eliminated, these quality differences could be limited by processing the serum before its direct use in culture or its fractionation. Although the centrifugation and dialyses have made most lots of human serum usable in culture, variation in quality still exists and is an area of future study. Some of the variation is because of quantitative differences in the composition of serum obtained from different donors. Figure 2 shows the difference between the elution profile of two different donors using the microbead column method for preparation of growth factors. These differences are consistent and can be duplicated when using aliquots of serum from the same donor. Their full significance remains to be determined.

Recent publications raise the question of using serum homologous with the cultured cell species and its possible affects. Choi et al [*11*] determined

TABLE 2—*Fetal calf serum, quality of units used in HED orofacial tissue culture.*

Lot	Date	Units	Source	Rejection Level Units	Rejection Level %	Rejection Level Cause
October	1977	2	A	0	0	...
December	1977	3	A	1	33	DCG[a]
January	1978	5	B	3	60	TCL[b]
March	1978	4	A	4	100	SED[c]
May	1978	8	A	6	75	TCL
October	1978	4	B	3	75	TCL
January	1979	4	B	2	50	DCG

[a] DCG = diminished cell growth.
[b] TCL = toxic with cell lysis.
[c] SED = excessive sedimentation.

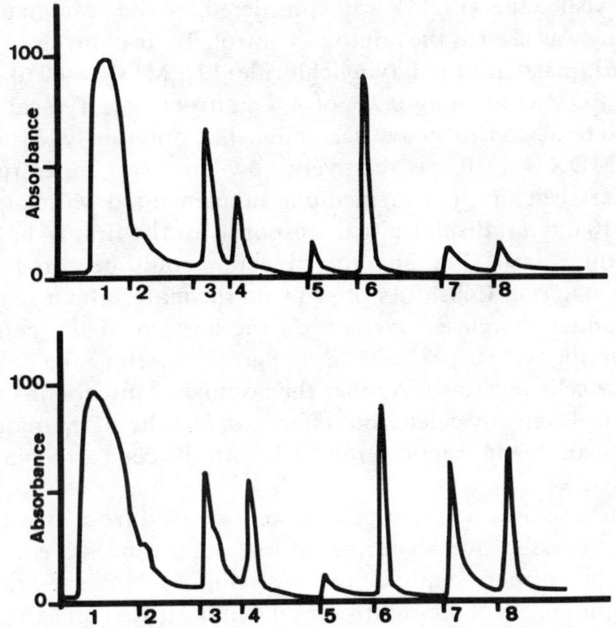

FIG. 2—*Elution profiles of two lots of serum from two individual donors carried out under identical conditions using a glass microbead column. Numbers 1 to 8 along the abscissa represent elution fractions. Compare the constancy of fraction numbers, their position, and the variation in peak heights.*

that human serum potentiated the growth of human chondrocytes in vitro. However, in this study no techniques other than centrifugation were used to remove potential toxicants. An investigation by Selten et al [*12*] has demonstrated the need of species specific serum factors. Their evidence has shown that a rat hepatoma cell line no longer produced detectable alpha-fetoprotein and albumin messenger ribonucleic acids (mRNAs) when grown in fetal calf serum. When rat serum was used as the medium supplement, the cell line resynthesized the appropriate mRNAs.

Although whole serum as a medium additive produces effective growth in vitro, and a homologous serum may produce fewer biochemical deviations than a heterologous one, an ideal situation would be the use of a serum fraction that produced no abnormal metabolic changes and for which specifications can be written. Thus far, the alpha-l-protein fractionated from human serum can support rapid growth of established cell lines and is therefore recommended as part of the test system. It is expected that current research will reveal serum fractions that can be used to grow primary or first passage cells for use in the test system. Although several fractions are in a testing phase, none are currently being used in this capacity.

Several runs of the optional testing methods were compared. In each case polydimethylsiloxane (PDM) was considered as the test material. Teflon FEP sheeting was used as the nontoxic control. Toxic controls included polyurethane [*6*], plasticized poly(vinylchloride) [*7*], MDX 4-4210 grade of silicone, and PDM containing 0.2% of 4,6-dinitrocresol. Typical reactions of the first two toxic controls have been published previously [*8*]. The toxic response of MDX 4-4210 was relatively slow. In a test series run using the Girardi heart cell line [*3*] in medium supplemented with alpha-protein, MDX 4-4210 did not display a toxic response for the first 24 h. However, at the end of three days, (Fig. 3), a toxic reaction could be noted in the center hole of the material. Cells did not grow on the material per se and actually grew best under the glass coverslip on the bottom of the petri dish. The farther from the test strip, the better the growth. After nine days in culture very few live cells remained. Neither the nontoxic control nor the test material PDM displayed any deleterious effects. In fact the PDM material (Fig. 4) supported a confluent monolayer growth with the cells attaching to the test material itself.

When a test series using first passage cells derived from skin explants was run using processed human serum, neither the nontoxic control nor the PDM specimen displayed a toxic response (Fig. 5). However, the toxic control, PDM plus 0.2% of 4,6-dinitrocresol demonstrated an extreme toxic response. Within 24 h cell death was evident, and by day four only cellular de-

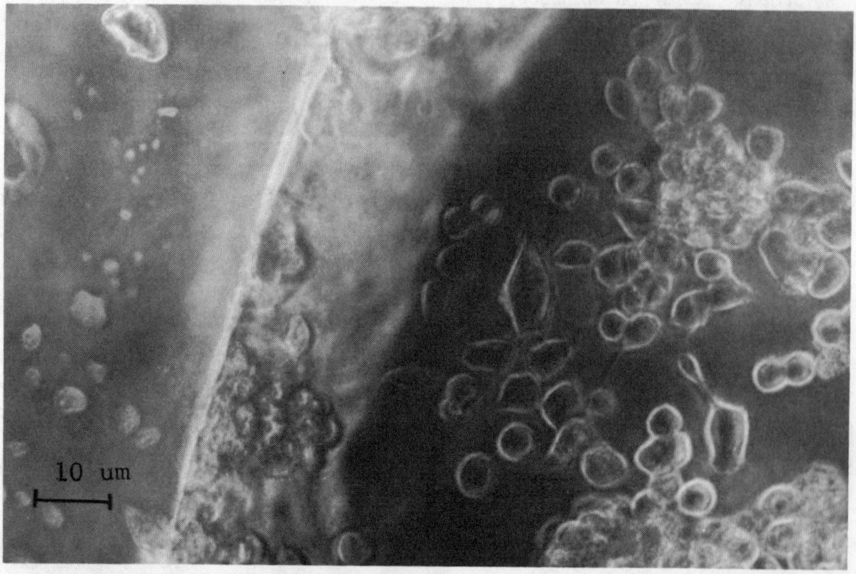

FIG. 3—*A toxic response of Girardi heart cells* [3] *to MDX 4-4210 grade silicone can be noted after three days of culture. Pictured is a section of the center hole of the test material (left) along which are dead clumped cells.*

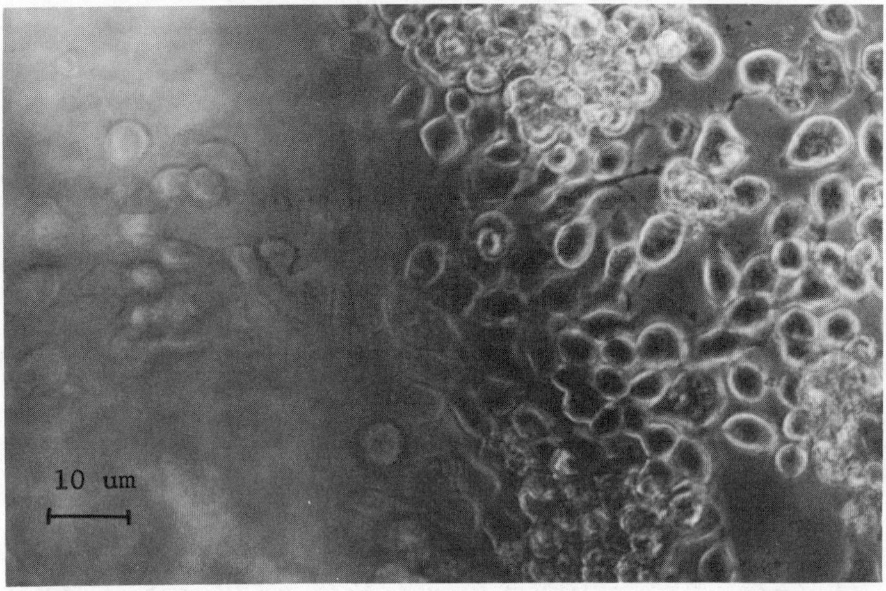

FIG. 4—*This photograph demonstrates the growth of Girardi heart cells* [3] *in a test culture containing polydimethylsiloxane. Cells are moving from the center hole onto the material itself* (*left*).

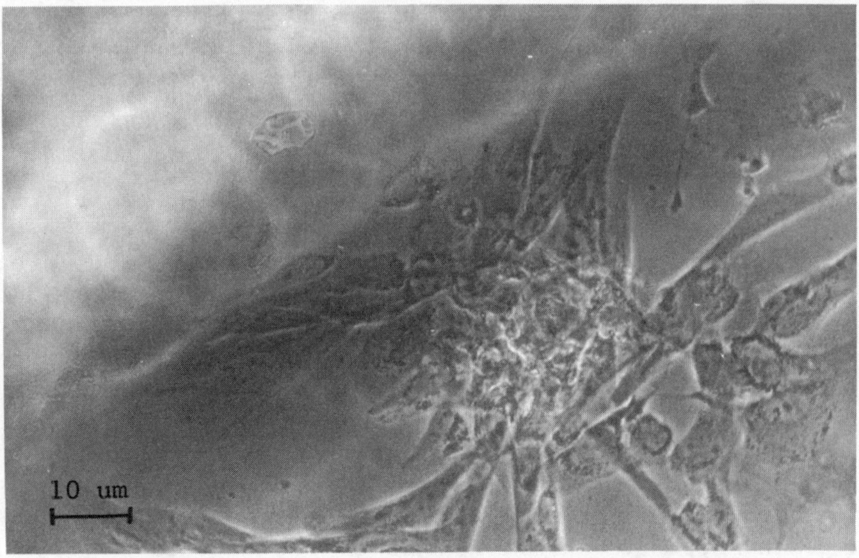

FIG. 5—*First passage skin cells were grown with the PDM material for six days in culture. Note that cells approach and grow on the material.*

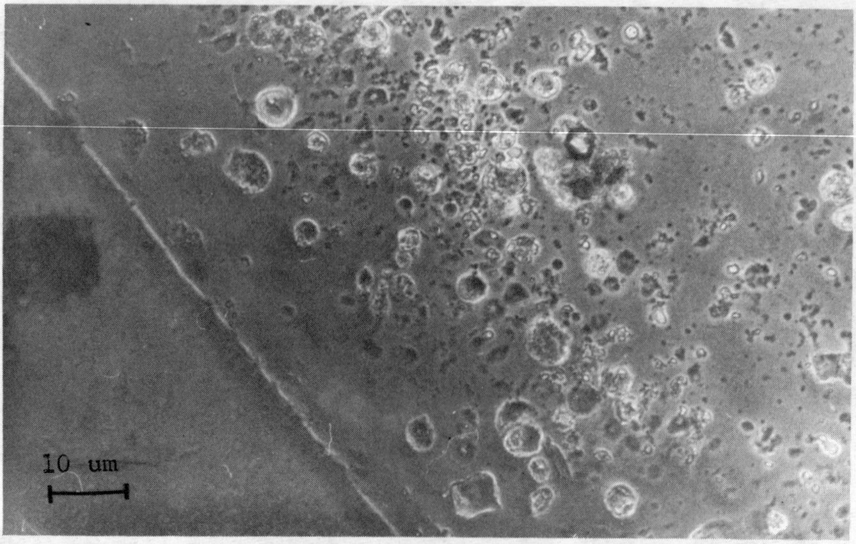

FIG. 6—*A maximum toxic response of first passage skin cells to PDM containing 0.2% of 4,6 dinitrocresol is demonstrated clearly. Only cellular debris remains after four days in culture.*

bris remained (Fig. 6). Although so far we have used this test to determine toxicity by evidence of cell death, in less evident toxic reactions cell growth determinations are recommended either by cell count or DNA analysis. Other nondetectable macroscopic changes may be monitored through assays of biochemical markers such as enzymes. These would be specific for cell types used. This test system is an attempt to duplicate in vitro as closely as possible some parameters of the in vivo system.

Summary

The biocompatibility test systems described in the report are presented on the assumption that human physiology can be reasonably approximated in tissue culture. This is accomplished by using homologous cells and serum. It is concluded that data obtained in this manner are more relevant to the human species than that obtained by heterologous test systems currently reported.

The tests provide a short term assessment of the safety of orofacial prosthetic materials. Both new biomaterials and those used for prolonged periods in reconstructions can be tested. Results from these test methods can be used to determine specifications and standards for safe and effective orofacial reconstructions.

References

[1] Holmes, R., Mercer, G., and Mohamed, N., *In Vitro,* Vol. 15, No. 7, 1979, pp. 522–530.

[2] Holmes, R., *Journal of Cell Biology,* Vol. 32, No. 2, 1967, pp. 297–308.

[3] Girardi, A. J., Warren, J., Goldman, C., and Jeffries, G., *Proceedings of the Society for Experimental Biology,* Vol. 98, No. 4222, Jan. 1958, p. 18.

[4] Nelson-Rees, W. A. and Flandermeyer, R. R., *Science,* Vol. 191, 1976, pp. 96–98.

[5] Barnes, D. and Sato, G., *Analytical Biochemistry,* Vol. 102, 1980, p. 255.

[6] Gonzalez, J. B., in *Proceedings,* First International Symposium on Facial Prosthetics, 19–23 April 1976, pp. 65–75.

[7] Robeson, L. M., Saunders, W. J., and Matzner, M., "Development of Improved Materials for Extraoral Maxillofacial Prosthesis," National Institute for Dental Research, Contract No. Nol-DE-42436, final technical report, 1976.

[8] Lontz, J. F. and Schweiger, J. W., *Bulletin of Prosthetics Research,* BPR 10-30, Fall 1978, pp. 223–244.

[9] Christensen, H. E., Ed., *The Toxic Substances List,* U.S. Department of Health, Education, and Welfare, Rockville, Md., June 1972, p. 209.

[10] Merchant, D. J., Kahn, R. H., and Murphy, W. H., *Handbook of Cell and Organ Culture,* Burgess Publishing Company, Minneapolis, Minn., 1964, pp. 176–187.

[11] Choi, Y. C., Morris, G. M., Lee, F. S., and Sokoloff, L., *Connective Tissue Research,* Vol. 7, No. 2, 1980, pp. 105–112.

[12] Selten, G. C. M., Selten-Versteegen, A. M. E., and Yap, S. H., *Biochemical and Biophysical Research Communications,* Vol. 103, No. 1, 1981, pp. 278–284.

Steven G. Meyers,[1] *Michael B. O'Connell,*[1] *and Michael S. Korczynski*[1]

Cell-Culture Cytotoxicity Testing of Candidate Device Materials

REFERENCE: Meyers, S. G., O'Connell, M. B., and Korczynski, M. S., **"Cell-Culture Cytotoxicity Testing of Candidate Device Materials,"** *Cell-Culture Test Methods, ASTM STP 810,* S. A. Brown, Ed., American Society for Testing and Materials, 1983, pp. 88–93.

ABSTRACT: In vitro cytotoxicity evaluation of candidate materials for medical devices is described. Materials and material extracts are added to freshly seeded human embryonic lung fibroblast (IMR-90) cell cultures. Both negative control and positive control materials are included in each evaluation of test materials. The negative control material does not cause detectable cytotoxicity to the cell system, and the positive control material causes cytotoxic effects.

Controls are used to monitor the cell system over time and detect any changes in the system that would alter the testing results. Results are scored on the basis of degree of morphological deterioration of cells and time required to obtain maximum deterioration. Candidate materials are evaluated comparatively in terms of their relative lack of cytotoxic effects on cell culture. Final results are not interpreted on a pass/fail basis but in terms of which materials are most likely to demonstrate biocompatibility in vivo.

This paper also reviews the need to develop specific cell-culture cytotoxicity assays on the basis of the intended end use of the material being evaluated. Some historical information concerning endothelial cell-culture systems and the need to develop a standardized cytotoxicity assay for blood contact materials is reviewed.

KEY WORDS: cell culture, cytotoxicity, materials evaluation

Historically, in vivo animal systems have been used for evaluation of materials for potential toxicity associated with direct and indirect contact with human tissues. In vivo tests are expensive and in some cases require long periods of time to obtain interpretable end results. Cell culture offers a means of supplementing information acquired by animal testing. Such methodology is amenable to evaluation of large numbers of materials. Materials that prove acceptable in cell-culture testing can then be further evaluated in animal systems.

[1] Section manager, director of sterilization and technology, and microbiologist, respectively, Hospital Products Division, Abbott Laboratories, North Chicago, Ill. 60064.

Methods

Cell-culture cytotoxicity methodology can be used as an indicator of in vivo biocompatibility (compatibility within intact human or animal tissue). It is amenable to concurrent evaluation of a number of materials and is less costly than toxicity testing with animals. Currently, the IMR-90 cell line of human embryonic lung fibroblasts is being used. Materials are evaluated by direct contact and by extraction procedures. In the direct contact method, test material is sterilized and incubated with culture cells. In the extraction procedure, separate preparations of candidate materials are placed in extractants (0.9% saline, 5% ethanol in 0.9% saline, water for injection, cottonseed, and fetal bovine serum). The extractions are carried out with a material/extractant ratio of 1/10 (w/u) and with temperature/time regima that are dependent upon the method of sterilization of the final product, temperature lability of the test material, and temperature lability of the extracting medium. The cell cultures are prepared by seeding about 1.0×10^6 cells/mL of fresh-medium to cell-culture plates. The materials or material extracts are added to the cell cultures immediately after the cells are seeded so that any potentially cytotoxic moiety would be present during the initial attachment and growth phase of the cell culture and during the time of normal confluent cell monolayer before culture senescence. Cultures are observed microscopically at one, four, and seven days. The criteria for scoring of cytotoxic effect is shown in Table 1. Cytotoxicity is rated on a scale from zero to four, with zero representing no cytotoxic effect and four representing complete destruction of the cell culture. The time required for development of the cytotoxic response is taken into account as well as the degree of the cytotoxic response. For example, a material eliciting a cytotoxicity of four in one day would be considered more cytotoxic than a material eliciting a cytotoxicity of four in seven days.

At this time, the cytotoxicity of a material is reported as the maximum tox-

TABLE 1—*Criteria for scoring cytotoxic effect on cell cultures.*

Cytotoxicity Score	Condition of Cell Culture
Negative (0)	polyhydral or fibroplastic cells with discrete intracytoplasmic granules, no cell lysis
Slightly cytotoxic (1+)	more than 20% of the cells are round, loosely attached and without intracytoplasmic granules, occasional lysed cells present
Definite cytotoxicity (2+)	more than 50% of the cells are round and devoid of intracytoplasmic granules, extensive cell lysis and empty areas between cells
High cytotoxicity (3+)	greater than 70% of the cell layer rounded or lysed or both, extensive empty areas
Extreme cytotoxicity (4+)	near complete destruction of the cell layer

icity observed in testing and the minimum amount of time to cause the response. The data are not handled on a pass/fail basis. Materials are ranked relative to their cytotoxic effect.

In all evaluations, control materials are utilized. A polyethylene reference material is used as a negative control; it does not cause an observable cytotoxic response in the IMR-90 cells. The second reference material is a polyvinyl chloride (PVC) with organotin; it typically causes a four cytotoxicity at about seven days. The purpose of the reference materials is to monitor consistency of the cell line with regard to its sensitivity to material toxicities. The reference materials are prepared in sufficient quantities to carry out a large number of tests over a long period of time. Before depletion of a given preparation of reference stock, a fresh stock is prepared; comparative evaluation of the new and old stock is carried out. The cytotoxic effect of the new reference stock must be the same as for the old stock before the new stock is incorporated as a control for evaluation of test materials.

Over the past two years, we have evaluated a variety of different materials for a number of different end uses. Materials that are intended for a given purpose have been compared with regard to their ranked cytotoxicity scores, and the better candidates have been selected for further evaluation by other means.

Thoughts for Future Consideration

Materials can be categorized based upon their end use and degree of intimacy with tissues. For example, the Health Industry Manufacturers Association (HIMA) has developed a testing panel based upon product end use, and this information is present in HIMA Report No. 78-7 [*1*]. Following the HIMA philosophy, we have asked ourselves whether a panel of cell-culture lines could be used to evaluate potential toxicity of materials. Such a panel would require different cell lines, each with physiological and morphological characteristics that reflect particular physiological and morphological characteristics of tissues in vivo. In this regard, fibroblastic cells lines, such as the IMR-90, are somewhat limited in that they lack much of the specialized morphology and function of cells within the specialized tissue of the body. However, advances in cell-culture technology have made it possible to culture cells on a relatively routine basis, which have characteristics that are more typical of some of the specialized cells in living tissues. From an industry view point, we would like to see such systems developed to a point where they could be used as a standard ASTM test. We would like to review some information that appears in the literature concerning a cell system that could be very useful in evaluating blood contact materials.

In 1921, Lewis [*2*] first observed the outgrowth of endothelial cells from explants of embryonic chick liver. Later, Woodward and Pomerate [*3*] developed a method for culturing capillaries from fragments of bone marrow.

By 1950, over 100 articles had been published related to the culture of endothelial cells and tissue [*4*].

In 1973 an important advance was made by Maruyama [*5*] who obtained an increased yield of endothelial cells by enzymatically dissociating the initial lining of human umbilical cord vessels. Maruyama used a trypsin solution at 37°C in his dissociation methods. He believed that the cytological and histochemical features of the isolated endothelial cells were not distinctive. Electron microscopy revealed that such endothelial cell cultures could be contaminated by vascular smooth muscle. Other investigators such as Fryer et al [*6*] confirmed Maruyama's findings. These findings revealed that cultured endothelial cells did not remain viable for a long time, and rapidly growing fibroblast-like contaminants frequently were present in the endothelial culture.

The earlier investigators were not aware that the contaminants in their endothelial cell cultures were probably providing growth factors to the endothelial cells. Growth factors and their effects upon endothelial cells were not well recognized until the early 1970s.

Recently, improved methods for the harvesting of endothelial cells have been used. Jaffe et al [7–9] and Gimbrane et al [*10–13*] used buffered collagenase instead of trypsin for harvesting endothelial cells from the intestinal lining of large blood vessels. These methods yielded homogeneous primary cultures of endothelial cells that were capable of being subcultured. Collagenase may permit selective digestion of the subendothelial basement membrane [*14*] of the intestinal lining of blood vessels. Thrombin (a serine protease) and a simple balance salt solution [*15, 16*] have been used successfully to release endothelial cells from blood vessels.

Relative to culture media, heterologous sera, including fetal calf, calf, and horse have been routinely used to supplement the media. However, relatively high serum concentrations (20 to 30%) appear necessary for optimal growth [*13*]. As a general rule, cell types that can receive nitrogenic stimulation from 10 to 20% serum are the easiest to cultivate [*6*]. Some of the more difficult to culture cells may require higher levels of factors found in serum, yet increased concentrations of serum may also contain cytotoxic factors that preclude cell development.

Thrombin has been shown to potentiate the growth response of certain cell types to epithelial growth factor (EGF), or fibroblast growth factor (FGF). This may be a result of enzymatic removal of cell surface proteins that uncover additional binding sites for growth factors.

Gospodariowicz et al [*17*] found bovine vascular endothelial cells derived from different vascular areas did not respond to EGF. The cells, however, responded well to FGF. The use of FGF has permitted the development of clonal endothelial cells of bovine origin. Human vascular endothelial cells are apparently responsive to both EGF and FGF, however, higher concentrations of FGF need to be used than with bovine cells. Using FGF and thrombin, Gospordarowicz et al found they could develop vascular endothelial cells capable of repeated passages with a high mitotic index.

While human vascular endothelial cells would represent ideal candidate cells, Gospodarowicz [*17*] in 1978 and Gimbrane [*6*] in 1976 have reported that human vascular endothelial cells have a limited life span in culture and require high amounts of human serum for maintenance. From a maintenance and perhaps cost standpoint, the use of human endothelial cells may have its limitations.

Wenzel [*18*] in 1980 and Acosta [*19*] in 1979 have utilized rat fetal heart cells, which may also serve as a model for endothelial cells. Acosta and Li [*1*] in 1979 also reported that rat heart endothelial cells release detectable quantities of lactate dehydrogenase (LDM) under conditions of hypoxia and glucose deprivation. This suggests an interesting method of possibly quantitating the cytotoxic effects of materials in contact with animal heart endothelial cells.

Summary

Currently, if one wishes to evaluate the biocompatibility of a blood contact material, a well documented and accepted in vitro cell or tissue-culture system does not exist that simulates the vascular system. A cell-culture system that mimics the vascular endothelial system would be ideal. Perhaps the ASTM F-4 Committee on Medical and Surgical Materials and Devices is an ideal technical group to bring some of these ideas to fruition.

References

[*1*] Health Industry Manufacturers Association, "Guideline for Evaluating the Safety of Materials Used in Medical Devices," HIMA Report No. 78-7, 1978, p. 7.

[*2*] Lewis, W. H., "Smooth Muscle and Endothelium in Tissue Cultures," *Anatomical Record,* Vol. 21, No. 72, 1921.

[*3*] Woodward, W. C. and Pomerat, C. M., "The Development of Patent Blood Vessels from Adult Human Rib Marrow in Tissue Culture," *Anatomical Record,* Vol. 117, 1953, pp. 663–683.

[*4*] Murray, M. A. and Kopech, G., *A Bibliography of the Research in Tissue Culture,* Vols. 1 and 2, Academic Press, New York, 1953.

[*5*] Maruyoma, Y., "The Human Endothelial Cell in Tissue Culture," *Zeitschrift Fuer Zellforschung und Mikroskopische Anatomie,* Vol. 60, 1973, pp. 69–79.

[*6*] Fryer, D. B., Girnbaum, G., and Luttrell, C. N., "Human Endothelium in Cell Culture," *Journal of Atherosclerosis Research,* Vol. 6, 1966, pp. 151–163.

[*7*] Jaffe, E. A., Nachman, R. L., and Becker, C. G., "Culture of Human Endothelial Cells Derived from Umbilical Cord Veins," *Journal of Clinical Investigation,* Vol. 51, 1972, p. 46a.

[*8*] Jaffe, E. A., Nachman, R. L., and Becker, C. G., "Culture of Human Endothelial Cells Derived from Human Umbilical Cord Veins," *Circulation,* Vol. 46, 1972, pp. 11–253.

[*9*] Jaffe, E. A., Nachman, R. L., and Becker, C. G., "Culture of Human Endothelial Cells Derived from Umbilical Veins, Identification by Morphological and Immulogic Criteria," *Journal of Clinical Investigation,* Vol. 42, 1973, pp. 2745–2756.

[*10*] Gimbrone, M. A., Jr., *Progress in Hemostasis and Thrombosis,* Vol. 3, 1976, pp. 1–28.

[*11*] Gimbrone, M. A., Jr. and Cotran, R. S., "Human Vascular Smooth Muscle in Culture: Growth and Ultrastructure," *Laboratory Investigation,* Vol. 33, 1975, pp. 16–27.

[12] Gimbrone, M. A., Jr., *Progress in Hemostasis and Thrombosis,* Vol. 3, 1976, pp. 1–28.
[13] Gimbrone, M. A., Jr., Cotran, R. S., and Folkman, J., "Human Vascular Endothelial Cells in Culture Growth and NDA Synthesis," *Journal of Cell Biology,* Vol. 60, 1974, pp. 673–684.
[14] Majno, G., "Two Endothelial Novelties, Endothelial Contraction; Collagenase Digestion of the Basement Membrane," *Thrombosis Dialthesis Haemorrhagica Supplement,* Vol. 40, 1970, pp. 23–30.
[15] Shepro, D., Rosenthal, M., and Balbouta, J., et al., "The Cultivation of Aorta Endothelium," *Anatomical Record,* Vol. 147, 1973, p. 523.
[16] Shepro, D., Batbouta, J. C., and Rubblee, L. S., "Serotonin Transport by Cultured Bovine Aortic Endothelium," Cir. Res., 36:799–806, (1975).
[17] Gospodarowicz, D., Greenburg, G., Bealecki, H., and Zetter, B. R., *In Vitro,* Vol. 14, 1978, pp. 85–118.
[18] Wenzel, D. G., Wheatley, J. W., and Byrd, G. D., *Toxicology and Applied Pharmacology,* Vol. 17, 1970, pp. 774–785.
[19] Acosta, D., and Li, C. P., *In Vitro,* Vol. 15, 1979, pp. 929–934.

Denis A. Magoffin,[1] *Debra L. Kushell,*[1] *and Richard J. Schlesinger*[1]

The Chromium-51 Release Cytotoxicity Test: Comparison to the Agar Overlay and Cell-Growth Inhibition Tests

REFERENCE: Magoffin, D. A., Kushell, D. L., and Schlesinger, R. J., **"The Chromium-51 Release Cytotoxicity Test: Comparison to the Agar Overlay and Cell-Growth Inhibition Tests,"** *Cell-Culture Test Methods, ASTM STP 810,* S. A. Brown, Ed., American Society for Testing and Materials, 1983, pp. 94–101.

ABSTRACT: The ^{51}Cr-release assay was compared to the agar overlay and cell-growth inhibition assays for cytotoxicity. L-929 mouse fibroblasts were used in each of the three assays. Incubation of Na $^{51}CrO_4$ labeled cells with increasing concentrations of phenol caused a dose-related accumulation of ^{51}Cr in the medium. The effective dose for a 50% response ED_{50} was 1.1% phenol. Increasing concentrations of phenol caused a dose-related increase in cell damage as determined in the agar overlay assay ($ED_{50} = 0.75\%$ phenol) and a dose-dependent inhibition of cell growth ($ED_{50} = 0.45\%$ phenol). The sensitivity of the cell-culture methods compared favorably with the in vivo intracutaneous reactivity test in rabbits. These studies demonstrate that the ^{51}Cr-release assay is of similar sensitivity to commonly used cell-culture cytotoxicity tests. The cell-culture tests offer better convenience and are faster and less costly than in vivo methods currently in use.

KEY WORDS: toxicity, cell culture, cytotoxicity, chromium release, agar overlay, cell growth inhibition

Plastic materials have been used extensively to fabricate a wide variety of medical devices varying from implantable prosthetic devices to transfusion sets and syringes. As a consequence, methods for testing the safety of plastics have been developed in order to minimize the risks to the patient. One of the most successful of the tests used to screen plastic materials used in medical practice is the technique of implanting small pieces of the plastic specimen into the paravertebral muscle of rabbits [*1–3*].[2] After several days, the im-

[1] Chairman of biology/molecular biology, microbiologist, and president, respectively, Bio-Technics Laboratories, Inc., 1127 Crenshaw Blvd., Los Angeles, Calif. 90019.

[2] The italic numbers in brackets refer to the list of references appended to this paper.

plantation sites are examined macroscopically and microscopically for signs of toxicity.

Use of the muscle implantation method involves considerable time and expense, especially when large numbers of specimens are to be screened. In recent years, tissue-culture methods have been adapted for use in routine cytotoxicity testing. Three of the most popular methods currently used are the agar overlay (agar diffusion) method [*4,5*], the chromium-release method [*6*], and the cell-growth inhibition technique [*7*]. The agar overlay method was found to be a highly sensitive indicator of toxicity when compared to in vivo testing methods [*8*].

In view of the widespread use of cell-culture cytotoxicity tests for the evaluation of new plastic materials, this study was performed to compare the chromium-release method to the agar overlay and cell-growth inhibition tests.

Procedure

Cell Culture

The L-929 mouse connective tissue cell line was propagated in a monolayer culture in Dulbecco's modified Eagle's minimal essential medium (MEM), supplemented with 10% fetal bovine serum, 2 m*M* *L*-glutamine and 100 μg/mL gentamicin at 37°C in 95% air/5% carbon dioxide. Cells were harvested with 0.25% weight by volume trypsin.

^{51}Cr-Release Assay

The chronium release assay was performed by the method of Spangberg [*6*]. Briefly, monolayers of cells were labeled by incubation with 1.5 μCi Na $^{51}CrO_4$ per 10^6 cells for 24 h. After the labeling period, cells were harvested with 0.25% trypsin and washed five times in medium containing 10 m*M* *N*-2-hydroxyethylpiperazine-*N*′-2-ethanesulfonic acid (HEPES). Labeled cells (10^6/dish) were incubated with increasing concentrations of phenol (0 to 6%) for 24 h in HEPES buffered culture medium. After incubation, the medium was collected, centrifuged (1000 $\times$ *g* for 10 min), and counted in a gamma spectrometer.

Agar Overlay Assay

The agar overlay assay was performed according to Rosenbluth et al [*4*] and Guess et al [*5*]. Monolayers of L-929 cells were washed with 3 mL phosphate buffered saline (PBS) (137 m*M* sodium chloride, 3 m*M* potassium sodium 10 m*M* phosphate chloride, pH 7.4). Two mL of agarose (0.7% in culture medium with 27 μg/mL neutral red) cooled to 40°C were added to each dish and allowed to harden. Increasing concentrations of phenol (0 to 6% final concentration) were pipetted onto the agar and the cells cultured for 24 h at 37°C in 95% air/5% carbon dioxide. After the incubation, the cultures were examined microscopically for uniformity of the monolayer, crenation, vacuolation, and cytolysis.

Cell Growth Inhibition Assay

Cell growth inhibition was measured by the method of Oyama and Eagle [*7*]. Cell monolayers were harvested with 0.25% trypsin and resuspended in culture medium. Cells (2×10^5) were incubated in 2 mL of increasing concentrations of phenol (0 to 6%) in 1300 mm culture tubes for 72 h at 37°C in 95% air/5% carbon dioxide. At zero time for one set of tubes and after 72 h for the test solutions, the cells are centrifuged, washed in PBS buffer, and the pellets assayed for protein by the method of Lowry et al [*9*].

Intracutaneous Reactivity

The intracutaneous reactivity test was performed according to the United States Pharmacopeia (USP) [*3*]. After clipping the hair from the backs of New Zealand white rabbits, the skin was swabbed with isopropyl alcohol. Increasing concentrations of phenol (0 to 6%; 0.2 mL) were injected intracutaneously at ten sites per concentration on the back of two rabbits. After 24 h, the injection sites were examined for erythema, eschar formation, edema, and necrosis (Table 1).

TABLE 1—*Phenol toxicity measured by the intracutaneous reactivity test.*

Phenol Concentration, % Weight by Volume	Skin Reaction
0.00	0
0.05	0
0.15	0.5
0.25	0.8
0.50	1.5
1.00	3.5
1.25	4
1.75	4
2.00	4
6.00	4

NOTE: Two rabbits were injected intracutaneously with increasing concentrations of phenol (0 to 6%) at ten sites per concentration. Skin reactions at each injection site were graded after 24 h as follows: 0, no erythema and no edema; 1, barely perceptible erythema and edema; 2, well-defined erythema and slight edema; 3, moderate to severe erythema and moderate edema; and 4, slight eschar formation and severe edema to necrosis. The results reported are the mean skin reaction at 24 h.

Statistical Analysis

Dose-response curves were analyzed and the ED_{50} calculated using the curve fitting program of Jaffe et al [*10*].

Results and Discussion

When L-929 cells labeled with $Na^{51}CrO_4$ were incubated with increasing concentrations of phenol, a dose-related increase in ^{51}Cr was observed in the medium during the incubation period (Fig. 1). Cells cultured in medium alone released 35% of the ^{51}Cr during the 24 h incubation. The maximum amount of ^{51}Cr released was 93% at 1.75% phenol or greater. The ED_{50} was $1.1 \pm 0.05\%$ phenol.

L-929 cells incubated with phenol using the agar overlay technique were damaged in a dose-related manner (Fig. 1) as determined by microscopic examination (Fig. 2) and supravital staining. In the absence of phenol, no cell damage was observed and the cells grew as a healthy monolayer (Fig. 2*a*). Increasing concentrations of phenol disrupted the monolayer, causing vacuolation, rounding up of the cells, detachment of the cells, and cytolysis (Fig. 2*b*). Concentrations of phenol greater than 1.75% caused maximum damage (>95%). The ED_{50} was $0.75 \pm 0.07\%$ phenol.

Increasing concentrations of phenol caused a dose-related inhibition of cell growth as measured by protein production (Fig. 1). Minimum and maximum effective doses of phenol were 0.15 and 1.0% phenol, respectively. The ED_{50} was $0.45 \pm 0.1\%$ phenol.

When the half maximum toxic doses of the three cell-culture cytotoxicity assays were compared, the three assays appeared to be approximately equally sensitive. The most sensitive assay (cell-growth inhibition) was only 2.4 times as sensitive as the least sensitive assay (^{51}Cr-release). These results were based on the toxicity of phenol. It is possible that different toxic substances may have different relative toxicities since each assay measures toxicity by a different biological endpoint. The ^{51}Cr-release assay measures the integrity of the plasma membrane of the labeled cells. When the integrity of the cellular membrane is disrupted, ^{51}Cr is released into the culture medium from the cytoplasm where it was bound [*6*]. The agar overlay assay relies on visual assessment of cellular damage [*4,5*] (Fig. 2), which can be aided by supravital staining. The cell growth inhibition assay measures cellular protein production as a measurement of cell growth [*7*]. These differences in end point are reflected in the differing shapes and slopes of the dose-response curves (Fig. 1).

We found that the dose-response curve for ^{51}Cr-release was shifted slightly to the right compared to those previously reported [*6*]. The difference may be in the substrain of cells used or subtle differences in the culture condi-

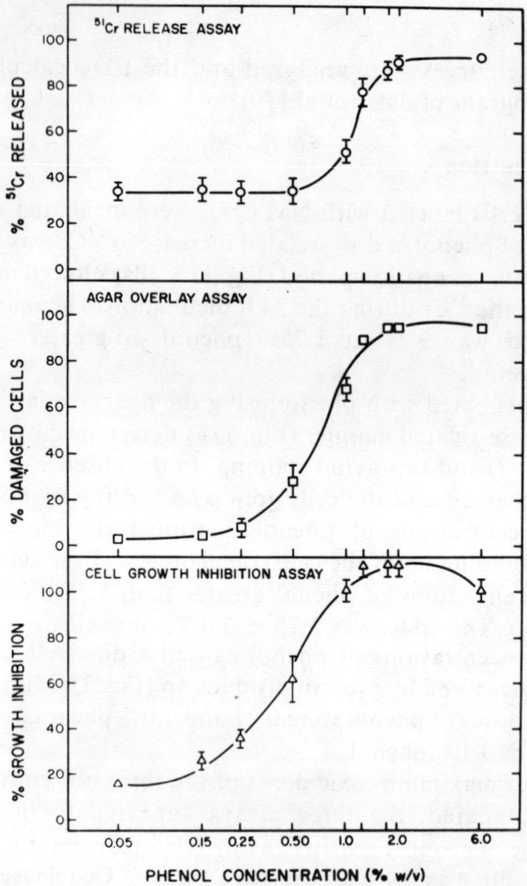

FIG. 1—*Effect of phenol on cultured L-929 mouse fibroblasts. L-929 cells were incubated with increasing concentrations of phenol. Toxicity was measured by ^{51}Cr released into the medium, observed cellular damage, and growth inhibition. The data are the mean ± standard error of three separate experiments with triplicate incubations per experiment.*

tions. We also found that the basal release of ^{51}Cr in control cultures was highly dependent on the washing procedure. Even with the utmost care in cell preparation and culture conditions, the basal release was subject to significant interassay variation. Care must be taken, therefore, to include appropriate controls in the assay protocol to ensure assay validity.

Each of the cell-culture methods have individual advantages. The ^{51}Cr-release and cell-growth inhibition assays have the advantage that the end point measurement yields hard numbers that minimize bias. The agar overlay method, even with supravital staining, is subject to significant interassay variation when being performed by technicians of varying experience. To a

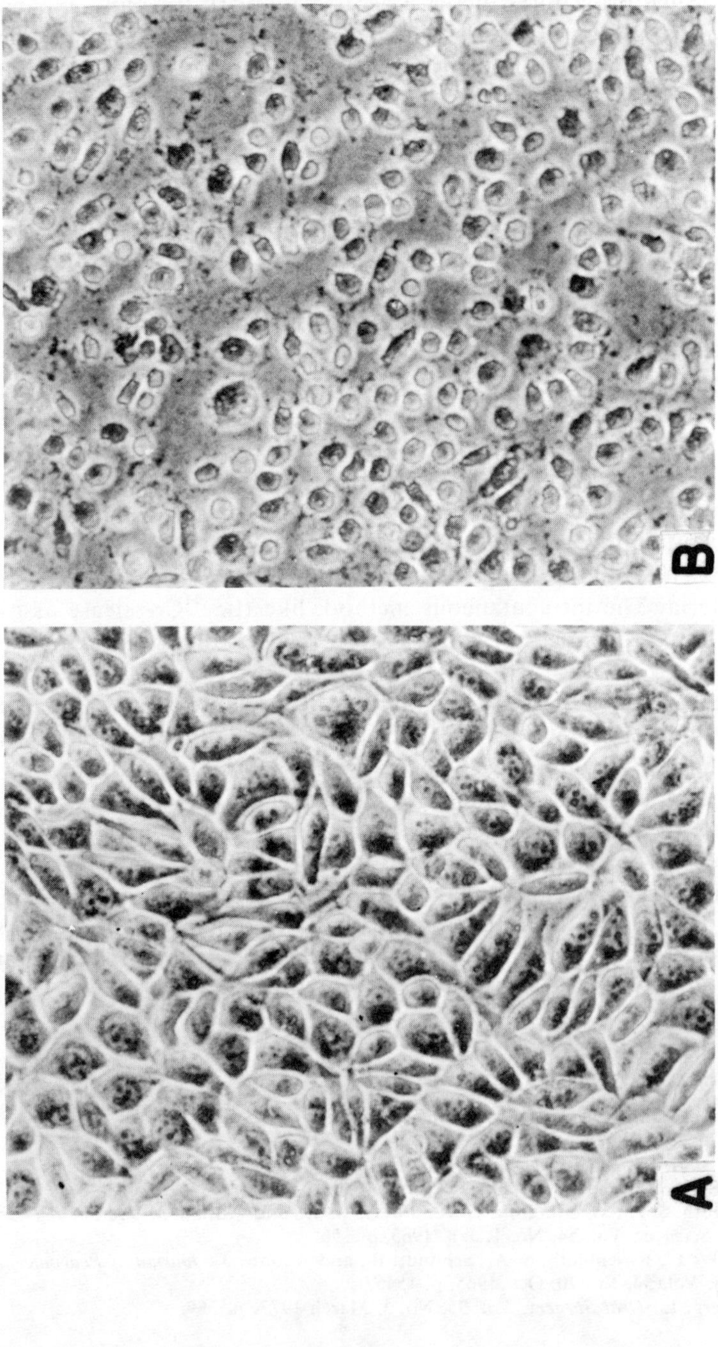

FIG. 2—*Effect of phenol on cultured cells in the agar overlay assay. Monolayers of L-929 cells were cultured in the* (A) *absence and* (B) *presence of 2% phenol for 24 h. The cultures were photographed using phase contrast microscopy* (×500).

large extent, bias can be minimized by using photographic records of the culture dishes and evaluation of the cultures by more than one experienced technician. Such procedures increase necessarily both time and expense of the assay. Most plastic materials, whether raw material or finished components, must be extracted by one of several methods such as those described in the USP [*11*]. This is true for the ^{51}Cr-release and cell-growth inhibition assays, although flat pieces of plastic cut to fit the culture dish can sometimes be used in the ^{51}Cr-release assay. The agar overlay assay has the advantage that pieces of any material, liquid, powder, or solid, regardless of shape, can be tested. It is therefore advantageous to have more than one of these assays available so that the appropriate assay(s) can be selected for each specimen.

The agar overlay assay has been shown to be highly sensitive for detection of toxicity relative to intramuscular implantation methods [*8*]. In order to compare the ^{51}Cr-release assay to in vivo methods, we tested phenol for intracutaneous reactivity. When increasing concentrations of phenol were injected intracutaneously into rabbits, a dose-related increase in skin reaction was observed (Table 1). The minimum, maximum, and half-maximum effective doses were 0.15, 1.25, and 0.7% phenol, respectively. The ED_{50} of the intracutaneous reactivity test was similar to the ED_{50} of the ^{51}Cr-release assay. This test is difficult to evaluate in a quantitative manner. Thus, small differences in toxicity could be completely masked because of the subjective evaluation criteria. The intracutaneous method, like the ^{51}Cr-release assay, is limited to liquid specimens or extracts of specimens but is considerably less convenient.

It is apparent from these studies that the ^{51}Cr-release assay compares favorably with other more routine cell-culture cytotoxicity tests. The potential exists, perhaps, by combining the agar overlay technique with the ^{51}Cr-release end point, for the development of a truly versatile cytotoxicity test that can be used for any type of specimen. When compared to in vivo tests, the cell-culture tests correlate well and offer considerable advantage in quantitation and the types of specimens that can be tested. In addition, the cell-culture tests require less technician time to perform and substantially less extensive support facilities. The cell-culture tests can therefore yield more quantitative data at about half the cost of in vivo tests.

References

[*1*] Brewer, J. H. and Bryant, H. H., *Journal of Pharmaceutical Sciences,* Vol. 49, No. 8, Aug. 1960, p. 652.

[*2*] Autian, J., *Artificial Organs,* Vol. 1, No. 1, Aug. 1977, p. 53.

[*3*] United States Pharmacopeia, 20th Revision, 1980, p. 952.

[*4*] Rosenbluth, S. A., Weddington, G. R., Guess, W. L., and Autian, J., *Journal of Pharmaceutical Sciences,* Vol. 54, No. 1, Jan. 1965, p. 156.

[*5*] Guess, W. L., Rosenbluth, S. A., Schmidt, B., and Autian, J., *Journal of Pharmaceutical Sciences,* Vol. 54, No. 10, Oct. 1965, p. 1545.

[*6*] Spangberg, L., *Oral Surgery,* Vol. 35, No. 3, March 1973, p. 389.

[7] Oyama, V. I. and Eagle, H., *Proceedings of the Society of Experimental Biology and Medicine,* Vol. 91, No. 2, Feb. 1956, p. 305.
[8] Pelling, D., Sharratt, M., and Hardy, J., *Food Cosmetic Toxicology,* Vol. 11, 1973, p. 69.
[9] Lowry, O. H., Rosenbrough, N. J., Fair, A. L., and Randall, R. J., *Journal of Biological Chemistry,* Vol. 193, No. 1, Nov. 1951, p. 265.
[10] Jaffe, M. L., Munson, P. J., and Rodbard, D., Four parameter logistic curve-fitting program for the TI-59 calculator, Biophysical Endocrinology Section, Endocrinology and Reproduction Research Branch, NICHHD, NIH, Bethesda, Md. 20014.
[11] United States Pharmacopeia, 20th Revision, 1980, p. 950.

Judith B. Ulreich[1] *and Milos Chvapil*[1]

In Vitro Toxicity Testing: A Quantitative Microassay

REFERENCE: Ulreich, J. B. and Chvapil, M., **"In Vitro Toxicity Testing: A Quantitative Microassay,"** *Cell-Culture Test Methods, ASTM STP 810,* S. A. Brown, Ed., American Society for Testing and Materials, 1983, pp. 102–113.

ABSTRACT: In vitro methods for testing the toxicity of biomaterials have many advantages over in vivo tests. They are more sensitive, simpler, more reproducible, rapid, inexpensive, easily duplicated, adaptable, and conservative of test materials. The microassay in use in our laboratory has the additional advantages of being quantitative, not requiring extensive microscopic observation or the presence of materials that may themselves be toxic (for example, agar and dyes), measuring parameters of cellular activity other than cell death, and being suitable for screening of a large number of specimens. The material to be evaluated, or in the case of insoluble materials an extract of the material, is incubated on monolayers of fibroblasts in 24 well culture plates. The activity of the fibroblasts is ascertained by measuring the rate of incorporation of ^{3}H-thymidine, $^{35}SO_4^=$, or ^{3}H-tryptophan. By proper choice of label, the effect of the biomaterial on cellular proliferation or production of macromolecules or both may be assessed. Materials that are cytotoxic cause inhibition of incorporation of radioactive label in the assay. Cell monolayers are also inspected microscopically for evidence of cytotoxicity. Examples are given to demonstrate use of the microassay.

KEY WORDS: biomaterials, cell culture, toxicity, microassay, in vitro, evaluation

In vitro methods for measuring the toxicity of biomaterials are becoming increasingly popular. They have several advantages over in vivo tests [*1–4*][2] including greater sensitivity, simplicity, reproducibility, rapidity, low cost, conservation of test materials, adaptability, and ease of duplication.

Several reported methods for testing biotoxicity in vitro involve microscopic examination of cells after exposure to various biomaterials [*3–6*]. The cells are fixed and stained [*6*], or dye exclusion is measured as an indication of cell viability [*3*]. A modification of the in vitro method using an agar plaque assay was introduced by Dulbecco [*7*] and further modified by Guess et al [*8*] who incorporated neutral red dye into the assay and Wennberg et al [*9*] who, in an attempt to reduce the discriminating effects of agar, assayed

[1] Research assistant, and professor and head of Section of Surgical Biology, respectively, University of Arizona Health Sciences Center, Tucson, AZ 85724.

[2] The italic numbers in brackets refer to the list of references appended to this paper.

succinate dehydrogenase activity in cells on a millipore filter resting on an agar support. Sudilovsky [*1*], noting the toxicity of neutral red itself, exposed cells to the dye for only short periods of time after exposure of the cells to test materials. The cell monolayer with agar overlay was then fixed with formaldehyde to improve the sensitivity of visual grading of results. Toxicity testing involving measurement of radioactive tracers seems to have been limited largely to measurement of release of ^{51}Cr from cells [*10*].

The majority of the assays described involve microscopic evaluation of the effects of toxic materials on cell monolayers. Some are complicated by the presence of materials (for example, agar and neutral red dye) that may themselves exert a discriminatory or toxic effect. The in vitro microassay system described here has the advantage of being quantitative, not requiring extensive microscopic observation or extraneous materials that might themselves be toxic, requiring few cells and small quantities of test material, measuring parameters of cellular activity other than death, and being suitable for easily screening large numbers of specimens.

Procedure

Cell Cultures

Fibroblasts (3T3) from the American Type Culture Collection were grown in 75-cm^2 Falcon® flasks in Eagle's minimum essential medium (MEM) with Earle's salts and L-glutamine, and 10% fetal-calf serum (FCS). The "complete" medium was supplemented with proline (11.5 mg/L), glycine (7.5 mg/L), ascorbic acid (50 mg/L), penicillin (100 units/mL), streptomycin (100 μg/mL), and Fungizone® (0.25 μg/mL). Cell cultures were maintained at 37°C in a humidified atmosphere with 5% carbon dioxide and 95% air. Monolayers were trypsinized (0.25%), counted in a hemocytometer, and plated in 24 well Linbro® plates for assays (Fig. 1).

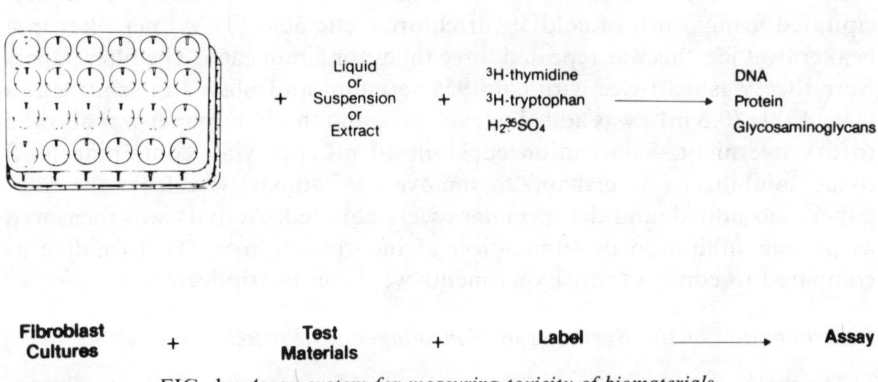

FIG. 1—*Assay system for measuring toxicity of biomaterials.*

Preparation of Extracts from Biomaterials

Test materials were washed, sterilized with ethylene oxide, and aired to remove remaining ethylene oxide. A measured amount of the material to be tested was cut into pieces approximately 5 mm in diameter and placed in capped plastic test tubes containing "complete" MEM to give 0.2-g biomaterial per 8-mL MEM. Extraction was carried out at 25°C for 24 h. (Extraction of materials designed for use in vivo can be done at 37°C, but if significant toxicity is caused by extracts prepared at 25°C, extraction at 37°C need not be done.) A series of extractions may be made to determine when maximal leaching out of toxic materials occurs. This is not necessary if a toxic response is observed in cells exposed to the 24-h extract. The extracts for use in cell culture were sterilized by filtration through 0.22 μm millipore filters and added to the cultures at various concentrations. As a control, MEM was allowed to stand at 25°C for 24 h. It is also possible to extract the material into water, MEM without serum, or other liquids.

Determination of the Incorporation of ^{3}H-Thymidine

Cells were plated in Linbro plates (4×10^4 cells per well) in "complete" medium and incubated for 24 h at 37°C under 5% carbon dioxide and 95% air. At 24 h, the test materials were added along with fresh "complete" MEM to a final volume of 1mL. To measure extract toxicity at a concentration of 50%, 0.5-mL extract is added to 0.5-mL "complete" MEM. A concentration of 10% combines 0.1 mL-extract, 0.9-mL "complete" MEM, and so forth. Tritiated thymidine (0.5 μCi/well-specific activity 2.0 Ci/mmol, New England Nuclear [NEN]) was added and the cells were incubated for an additional 24 h.

After 24 h the medium was removed and the cells washed twice with Hanks' balanced salt solution (HBSS). Distilled water (1 mL) was added to each well and the specimens sonicated for 15 s each. Aliquots of 100-μL from each well were spotted on Whatman® 3-MM filter paper. The DNA was precipitated using 3 mL of cold 5% trichloroacetic acid (TCA) per filter in a beaker over ice; this was repeated three times for 5 min each. The filter papers were then washed twice with cold 95% ethanol and placed in scintillation vials. Ether (0.5 mL) was added to each vial, and the filter paper was allowed to dry overnight. Scintillation cocktail, 10 mL per vial, containing NCS tissue solubilizer (Amersham) to remove the radioactivity from the filter paper, was added, and the specimens were counted. Activity was measured as percent inhibition or stimulation of incorporation of ^{3}H-thymidine as compared to control cells. Experiments were run in triplicate.

Determination of the Synthesis of Noncollagenous Protein

The method is similar to the above protocol for determining deoxyribonucleic acid (DNA) synthesis with the exception that 1 μCi of ^{3}H-tryptophan (specific activity 18.1 Ci/mmol [NEN]) is added (instead of ^{3}H-thymidine) to each microwell along with the test materials. Incorporation of label into TCA precipitable material from the cell homogenate is measured.

Determination of Glycosaminoglycan (GAG) Synthesis by Incorporation of $^{35}SO_4^=$

Cells were plated at 4 × 10^4 cells/well (for log phase studies) or 10^5 cells/well (for confluent phase studies) in 24 well Linbro plates and were incubated in "complete" medium for 24 h under 5% carbon dioxide and 95% air at 37°C. The medium was replaced with test materials and fresh complete sulfate-free MEM to make 1 mL along with 20 μCi $H_2{}^{35}SO_4^=$ (specific activity 43 Ci/mg [NEN]). The medium from each well was harvested after 24 h and dialyzed against 0.1*M* ammonium sulfate for 4 h in the cold. Dialysis was continued for 24 h with four changes of cold water. The contents of each dialysis bag were placed in a scintillation vial with 10-mL Aquasol® and counted. After removal of the medium from the cells, the monolayer was rinsed twice with cold HBSS, pH 7.0. After addition of 0.5-mL distilled water per well, the contents of each well were sonicated for 20 s and assayed for protein content by the Lowry method [*11*]. Experiments were done in triplicate.

Microscopic Observation

Fibroblasts were plated in "complete" MEM at 4 × 10^4 cells/well in 24 well Linbro plates with 12-mm glass coverslips in each well (sterilized by soaking in ethanol and flaming) and incubated as described. At 24 h, extracts of sterile test materials were added along with fresh "complete" MEM to make 1 mL. At the desired times coverslips were removed, washed four times with phosphate buffered saline (37°C), fixed in 95% methanol for 1 min, air dried and stained with hematoxylin and eosin, and examined by light microscopy for signs of cellular alterations. Coverslips were rated according to degree of cellular rounding and number of cells present. Correlation of morphological alterations in cells exposed to toxic materials with the results of our microassay has been published previously [*12*].

Results

Examples from our research are shown to demonstrate the use of the in vitro microassay. Incorporation of ^{3}H-thymidine, $^{35}SO_4^=$, and ^{3}H-tryptophan into fibroblasts is measured in the presence of biomaterials or extracts of biomaterials in order to assess their possible toxicity in a quantitative manner.

Table 1 shows the results of an in vitro assay for evaluation of the cytocompatibility of various polymers by measurement of incorporation of ^{3}H-thymidine into 3T3 fibroblasts exposed to extracts of sponges made of var-

TABLE 1—*Toxicity of various sponges in tissue culture.*

Type of Sponge	^{3}H-Thymidine Incorporation with 50% Extract, % of control
Polyurethane	1[a]
Acetylcellulose	0[a]
Polyvinylalcohol	2[a]
Collagen tanned with formaldehyde	52[a]
Collagen tanned with isocyanate	78[b]
Collagen tanned with glutaraldehyde	71[a]
Collagen tanned with glutaraldehyde and methylated	88[c]
Collagen tanned with glutaraldehyde and washed with phosphate buffer	82[b,c]

[a] Significantly different from controls at $P < 0.001$ using Student's t-test.
[b] Significantly different from controls at $P < 0.01$ using Student's t-test.
[c] Significantly different from collagen sponge tanned with glutaraldehyde at $P < 0.05$ using Student's t-test.

ious polymers or treated with different cross-linking agents [*12*]. Several sponges of different composition were extracted into complete MEM as described and applied to log phase cells with fresh medium in a 1:1 ratio, the final concentration of test extract being 50%. The data indicate three levels of cytotoxicity. Polyurethane, acetylcellulose, and polyvinylalcohol sponge extracts are all extremely toxic to fibroblasts in our system. They all cause nearly 100% inhibition of incorporation of ^{3}H-thymidine into DNA. Of the three collagen sponges tested, isocyanate tanning appears to be the least detrimental in terms of biotoxicity with only 22% less thymidine incorporated than in control cells. However, with respect to the glutaraldehyde tanned collagen sponge, the results demonstrate that it is possible to reduce the toxic effect by either methylation or washing in neutral-pH phosphate-buffered saline. We have shown previously that collagen tanned with glutaraldehyde will retain the glutaraldehyde not only as a part of the cross-link with the collagen molecule but possibly also as a polymer independent of, or adsorbed to, the collagen matrix [*13*]. Prewashing the sponge in phosphate buffered saline probably removes some of the glutaraldehyde, and esterification methylation provides reduced extractability of the glutaraldehyde. It is clear that the in vitro method enables one to choose a sponge for in vivo implantation, which will exhibit the least toxicity with respect to cellular proliferation. In our laboratories we have done in vivo implantation tests in rabbits that confirm the relative degrees of toxicity of different sponges that we observe in vitro (D. P. Speer and M. Chvapil, unpublished results).

In a subsequent assay polyurethane sponges prepared by several different methods (known only by the manufacturer) were compared (Table 2) to determine whether the toxicity seen in our original assay could be reduced. Clearly, the different treatments used in preparation of these sponges re-

TABLE 2—*Incorporation of ^{3}H-thymidine by 3T3 fibroblasts exposed to extracts of various polyurethane sponges.*

Extract	^{3}H-Thymidine Incorporation, cpm			
	10% Extract,[a] mean ± SEM	Percent of Control	70% Extract, mean ± SEM	Percent of Control
None (control)	16713 ± 526	...	16713 ± 526	...
Sponge A	15130 ± 1046	90	14387 ± 295[b]	86
Sponge B	16290 ± 173	97	11211 ± 632[b]	67
Sponge C	16907 ± 397	101	7860 ± 771[b]	47

[a] SEM = standard error of mean.
[b] Significantly different from control at $P < 0.02$ using Student's t-test.

duced the toxicity significantly. Note that although no toxicity is apparent at 10% sponge extract, there is significant toxicity at 70% extract.

After noting the toxicity demonstrated by the glutaraldehyde tanned collagen sponge, we tested the effect of pure glutaraldehyde on fibroblasts using the same assay. Results presented in Fig. 2 clearly show that glutaraldehyde at concentrations of approximately 3 ppm and above causes almost 100% inhibition of incorporation of ^{3}H-thymidine into the fibroblasts. Leaching of glutaraldehyde from the cross-linked sponge may therefore account for its toxicity.

Another study was done to determine whether the type of tube used to col-

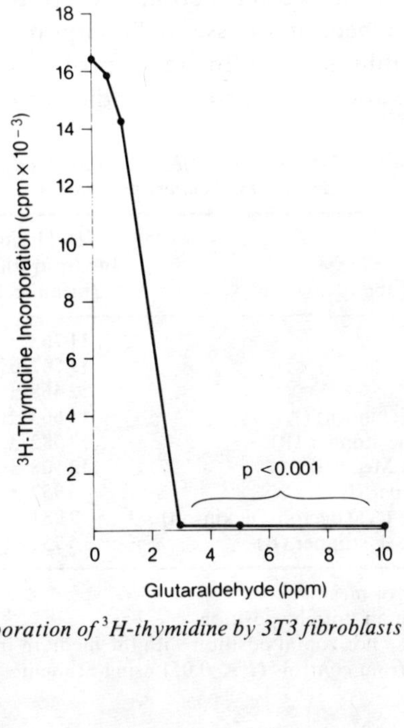

FIG. 2—*Inhibition of incorporation of ^{3}H-thymidine by 3T3 fibroblasts exposed to glutaraldehyde.*

lect blood or store human serum might affect results of in vitro experiments requiring blood drawn into such tubes [*14*]. In order to test tubes for contaminants that might alter cellular activity, a variety of commercially available tubes used for collection of blood and storage of serum were shaken while containing culture medium with FCS. The medium was then applied to 3T3 fibroblasts in culture. Results of the in vitro analysis of tubes preincubated with medium containing 10% FCS are shown in Table 3. The disposable glass tube and the siliconized tubes in an upright position had little effect on the incorporation of thymidine into DNA in fibroblasts. The serum separating tube (SST), which contains a "plug" of material at its base allowing better separation of serum from cells upon centrifugation, inhibited incorporation of label by 15%. It represented the only tube showing a toxic effect when tested in an upright position. When medium was incubated with tubes shaken in a horizontal position, red stoppered tubes caused 41% inhibition of thymidine incorporation, and variegated black and red stoppered tubes (SST) caused 51% inhibition. The blue stoppered tubes (which were developed for use in clinical chemistry laboratories where other tubes had been shown to cause interference in some assays) caused no toxic effect. Parafilm®, when tested with glass tubes, also failed to be responsible for any toxic effects in the fibroblasts.

Leaching of some substances, mainly zinc and copper, into serum collected in Vacutainer® tubes (Becton Dickinson) has been reported by several authors [*15–18*]. Based on our previous experience [*19,20*], zinc ions per se could be inhibitory to various cell functions. Alternatively, the tubes or caps could be depleting the medium of essential components by adsorption. Our finding of significant inhibition of fibroblast proliferation when the medium washed only the tube and not the stopper indicates clearly that the plug in

TABLE 3—*Incorporation of ^{3}H-thymidine by fibroblasts exposed to medium preincubated in various blood collection tubes.*

Tube	^{3}H-Thymidine Incorporation, cpm (mean ± SEM)	Percent of Control
None (control)	11 763 ± 443	...
Glass (A)[b]	11 567 ± 304	98
Glass + parafilm (B)[c]	12 468 ± 192	106
Siliconized (Becton Dickinson) (A)	11 663 ± 261	99
Siliconized + dark blue stopper (B)	12 582 ± 277	107
Siliconized (Sherwood Medical) (A)	12 108 ± 368	103
Siliconized + red stopper (B)	6 957 ± 190[d]	59
Serum separating tube (A) (Becton Dickinson)	9 981 ± 257[d]	85
Serum separating tube + stopper (B)	5 751 ± 215[d]	49

[a] SEM = standard error of mean.
[b] A tubes were shaken in an upright position.
[c] B tubes were shaken in a horizontal position with the medium in contact with the stoppers.
[d] Significantly different from controls ($P < 0.05$) using Student's t-test.

the serum separating tube contains certain substance(s) that can adversely affect cellular activity.

Since the 1940's, organosilicon compounds and polymers have been used widely in medicine. In order to determine the relative toxicities of several such compounds, they were diluted in dimethyl sulfoxide (DMSO) and applied to 3T3 fibroblasts (Table 4). Octamethyltetrasiloxane, the compound with the greatest number of silicon atoms per molecule, was further evaluated using the microassays for ^{3}H-thymidine, $^{35}SO_4^=$, and ^{3}H-tryptophan incorporation (Table 5). The degree of incorporation of tracers in the control cultures as shown here is typical for these cells. DMSO used for dilution of organic compounds does not cause cytotoxicity if it is limited to 0.5% (5 μL DMSO/mL medium). Several possible activities of silicon in connective tissue structure and metabolism have been postulated.[3] These results demonstrate that this organosilicon compound inhibits DNA and protein synthesis in log phase cells and stimulates the production of glycosaminoglycans in confluent cells. The stimulation of glycosaminoglycans production in confluent cells (which are always more resistant to the effects of toxic materials than log phase cells) may represent a reaction of the cells to a toxic substance. Incorporation of $^{35}SO_4^=$ is a valuable tool for measuring the response of cells to various materials, and incorporation of label often will be inhibited by toxic materials, especially when measured in log phase cells.

Because an established cell line that is relatively stable over time is used in

TABLE 4—*The effect of six organosilicon compounds on the incorporation of ^{3}H-thymidine by 3T3 fibroblasts.*

Compound[a]	Silicon atoms/molecule	^{3}H-Thymidine Incorporation, cpm (mean ± SEM)[b]	^{3}H-Thymidine Incorporation, % of control
Control	...	9616 ± 247	...
DMSO control	...	9510 ± 335	...
Phenyltriethoxysilane	1	1471 ± 290[c]	15
Tetraethylorthosilicate	1	2587 ± 663[c]	27
Dodecoxytriethoxysilane	1	9252 ± 360	97
3,3,5-Trimethyl-1,1-dimethoxy-1-sila-2,6-dioxacyclohexane	1	9399 ± 247	99
2-Methyl-2,4-di (trimethoxysiloxy) pentane	2	8502 ± 427	89
Octamethyltetrasiloxane	4	7254 ± 737[d]	76

[a] DMSO diluted silicon-containing compound (5 μM) incubated on log 3T3 fibroblasts.
[b] SEM = standard error of mean.
[c] Significantly different from DMSO controls ($P < 0.001$) by Student's t-test.
[d] Significantly different from DMSO controls ($P < 0.05$) by Student's t-test.

[3] Ulreich, J. B., Tillema, L. M., and Chvapil, M., "The Effect of Organosilicon Compounds on Fibroblast Activity in Vitro," unpublished data.

TABLE 5—*The effect of octamethyltetrasiloxane on the incorporation of ^{3}H-thymidine, $^{35}SO_4^=$, and ^{3}H-tryptophan by 3T3 fibroblasts.*

Additions to the Medium	^{3}H-Thymidine Incorporation in Log Phase Cells, cpm (mean ± SEM)[a]	$^{35}SO_4^=$ Incorporation in Confluent Cells, cpm/μg protein (mean ± SEM)	^{3}H-Tryptophan Incorporation in Log Phase Cells, cpm (mean ± SEM)
None (control)	15204 ± 106	928 ± 35	1191 ± 39
DMSO (control)	14348 ± 217	935 ± 41	1161 ± 41
Octamethyltetrasiloxane (diluted in DMSO)	...	...	...
10 μM	14796 ± 199	993 ± 33	1163 ± 4
25 μM	15044 ± 117	979 ± 13	1206 ± 25
50 μM	5323 ± 279[b]	1017 ± 74[b]	599 ± 4[b]
75 μM	2875 ± 783[b]	948 ± 51	480 ± 69[c]
100 μM	2297 ± 571[b]	1115 ± 37[d]	316 ± 32[b]
200 μM	152 ± 17[b]	2153 ± 44[b]	49 ± 7[b]

[a] SEM = standard error of mean.
[b] Significantly different from DMSO controls ($P < 0.001$) by Student's t-test.
[c] Significantly different from DMSO controls ($P < 0.01$) by Student's t-test.
[d] Significantly different from DMSO controls ($P < 0.02$) by Student's t-test.

these microassays, and because cells are counted before plating, inter-assay variation is minimal. The standard error values presented in the preceding tables demonstrate the magnitude of inter-assay variation.

Discussion

For several years there have been reports in the literature of a wide variety of cell-culture techniques to measure the toxicity of biomaterials [*21,22*]. The results reported here show that incorporation of radioactive precursors into cells in microwell assays may be valuable for more precise quantitation of results in such experiments and, because extensive microscopic observation is not necessary, they are rapid. This assay obviates the necessity of using dye exclusion, sometimes tested in conjunction with the agar overlay technique, as a measure of cell death or membrane damage. These methods have recently come under attack as reliable methods of determining cell survival [*23,24*].

Screening of biomaterials using a variety of tests involving both in vivo long-term animal studies and several in vitro methods would be extremely valuable in the assessment of any material intended for medical use. Thus, extracts of biomaterials that demonstrate toxicity in the in vitro microassay can further be evaluated using gas liquid chromatography and mass spectrometry. For in vivo testing, biomaterials can be implanted subcutaneously or in rabbit vaginas for ten days. The reaction of the mucosal epithelium can then be measured since the vaginal epithelium is sensitive both to toxic substances leaching from the biomaterial and to the surface morphology of the implant. However, for initial screening of large numbers of specimens, the in vitro assays presented here have many advantages and are adaptable to many situations. The assays are more economical, faster, more conservative of test materials, and more sensitive than animal tests. In addition, many parameters of the in vitro assay can be altered easily. The time of incubation, the type of radioactive label, the concentration of test material, the cell type, the phase of cell growth (log or confluent), the serum concentration in the medium, and the type of culture medium can be modified to suit the individual experiment. Assays can be carried out on glass coverslips in the microwells if scanning electron microscopy or staining and examination under light microscopy is desired.

The assay can be used to measure the effect on cells of liquids (which can be diluted in water or culture medium and added directly to the cultures if they are compatible with the aqueous environment or diluted in DMSO if they are nonpolar) or solids (some of which can be dissolved either in water, culture medium, or DMSO and added to the culture medium; others that can be suspended in the medium, and others that can be extracted into water or culture medium at room temperature and the extracts tested). Testing extracts of solids in vitro is preferable to testing solids themselves because exact amounts of extract can be added to the cells in replicate without concern about the size or shape of the solid, any response from the cells caused by insufficient supplies of gas and nutrient [*25*], or any decrease in sensitivity caused by an agar overlay that might absorb components and impede diffusion of toxic substances [*2*].

Previous in vitro assays for toxicity that use radioactive materials have generally been limited to measuring the release of ^{51}Cr, and this release has been shown to be insensitive in some assays [*23*]. If ^{51}Cr is covalently bound to basic amino acids of intracellular proteins that leak out of the cell as a consequence of membrane damage [*26*], ^{51}Cr release might not occur unless the cell were killed or its membrane damaged extensively. By measuring parameters of cell metabolism as well as proliferation, it is possible to detect toxicity caused by materials that do not cause cell death as assessed by ^{51}Cr release. However, some materials tested may induce a delay in cell cycle progression, and this lag in the multiplication rate of the cells may produce results that cannot be distinguished from those showing a decreased number of cells caused by cytotoxicity. For this reason, microscopic observation of cells during the assay is essential.

In summary, because of the cumbersome nature and relative lack of sensitivity and reproducibility of some in vitro culture methods currently used for assessing the toxicity of biomaterials, we believe that measurement of incorporation of radioactive precursors into DNA, GAG, or protein in cells in culture is a valuable improvement in the field of in vitro biotoxicity testing.

It is rapid, sensitive, quantitative, and can be used to determine the effects of materials used for medical purposes in vivo on parameters of in vitro cellular activity other than death.

Acknowledgments

We wish to thank John Wiley and Sons, Inc. for permission to reproduce Figs. 1 and 2 and Table 1 from our recent publication [*12*].

References

[*1*] Sudilovsky, O., *Tissue Culture Association Manual,* Vol. 3, No. 2, Aug. 1977, pp. 607–612.
[*2*] Wilsnack, R. E., Meyer, F. J., and Smith, J. G., *Biomaterials, Medical Devices and Artificial Organs,* Vol. 1, No. 3, Oct. 1973, pp. 543–562.
[*3*] Rice, R. M., Hegyeli, A. F., Gourlay, S. J., Wade, C. W. R., Dillon, J. G., Jaffe, H. and Kulkarni, R. K., *Journal of Biomedical Materials Research,* Vol. 12, No. 1, Jan. 1978, pp. 43–54.
[*4*] Rosenbluth, S. A., Weddington, G. R., Guess, W. L., and Autian, J., *Journal of Pharmaceutical Sciences,* Vol. 54, No. 1, Jan. 1965, pp. 156–159.
[*5*] Kawahara, H., Yamagami, A., and Nakamura, M., *International Dental Journal,* Vol. 18, No. 1, March 1968, pp. 443–467.
[*6*] Wilsnack, R. E., *Biomaterials Medical Devices and Artificial Organs,* Vol. 4, No. 3/4, June 1976, pp. 235–261.
[*7*] Dulbecco, R., *Proceedings of the National Academy of Sciences,* Vol. 38, No. 8, Aug. 1952, pp. 747–752.
[*8*] Guess, W. L., Rosenbluth, S. A., Schmidt, B., and Autian, J., *Journal of Pharmaceutical Sciences,* Vol. 54, No. 10, Oct. 1965, pp. 1545–1547.
[*9*] Wennburg, A., Hasselgren, G., and Tronstad, L., *Journal Biomedical Materials Research,* Vol. 13, No. 1, Jan. 1979, pp. 109–120.
[*10*] Spangberg, L., *Oral Surgery, Oral Medicine and Oral Pathology,* Vol. 35, No. 3, March 1973, pp. 389–401.
[*11*] Lowry, O. H., Rosebrough, N. J., Farr, A. L., and Randall, R. J., *Journal of Biological Chemistry,* Vol. 193, No. 1, Nov. 1951, pp. 265–275.
[*12*] Ulreich, J. B. and Chvapil, M., *Journal Biomedical Materials Research,* Vol. 15, No. 6, Nov. 1981, pp. 913–922.
[*13*] Speer, D. P., Chvapil, M., Eskelson, C. D., and Ulreich, J. B., *Journal of Biomedical Materials Research,* Vol. 14, No. 6, Nov. 1980, pp. 753–764.
[*14*] Ulreich, J. B. and Chvapil, M., *In Vitro,* Vol. 18, No. 2, Feb. 1982, pp. 117–121.
[*15*] Cerklewski, F. L., *Nutrition Reports International,* Vol. 18, No. 1, July 1978, pp. 1–4.
[*16*] Guillard, O., Piriou, A., Mura, P. and Reiss, D., *Clinical Chemistry,* Vol. 25, No. 10, Oct. 1979, p. 1867.
[*17*] Handy, R. W., *Clinical Chemistry,* Vol. 25, No. 1, Jan. 1979, p. 197.
[*18*] Reimold, E. W. and Besch, D. J., *Clinical Chemistry,* Vol. 24, No. 4, April 1978, p. 675.
[*19*] Chvapil, M., Stankova, L., Zukoski, C., IV, and Zukoski, C., III, *Journal of Laboratory and Clinical Medicine,* Vol. 89, No. 1, Jan. 1977, pp. 135–146.
[*20*] Chvapil, M., Stankova, L., Bernhard, D. S., Weldy, P. L., Carlson, E. C., and Campbell, J. B., *Infection and Immunity,* Vol. 16, No. 1, April 1977, pp. 367–373.
[*21*] Autian, J., in *Critical Reviews in Toxicology,* L. Goldberg, Ed., The Chemical Rubber Co., Cleveland, 1972, pp. 1–40.
[*22*] Eagle, H. and Foley, G. E., *American Journal of Medicine,* Vol. 21, No. 5, Nov. 1956, pp. 739–749.

[23] Roper, P. R. and Drewinko, B., *Cancer Research,* Vol. 36, No. 7, July 1976, pp. 2182–2188.
[24] Yuhas, J. M., Toya, R. E., and Pazmino, N. M., *Journal of the National Cancer Institute,* Vol. 53, No. 2, Aug. 1974, pp. 465–468.
[25] Nakamura, M. and Thonard, J. C., *Journal of Dental Research,* Vol. 52, No. 6, June 1973, p. 1341.
[26] Ronai, P. M., *Blood,* Vol. 33, No. 2, Feb. 1969, pp. 408–413.

Robert J. Hay[1]

Availability and Standardization of Cell Lines at the American Type Culture Collection: Current Status and Prospects for the Future

REFERENCE: Hay, R. J., **"Availability and Standardization of Cell Lines at the American Type Culture Collection: Current Status and Prospects for the Future,"** *Cell-Culture Test Methods, STP 810,* S. A. Brown, Ed., American Society for Testing and Materials, 1983, pp. 114–126.

ABSTRACT: Procedures in use at the American Type Culture Collection (ATCC) for standardization of cell-culture lines are summarized with illustrations. Comments on the total holdings and distribution are included with a listing and brief description of representative cell lines that may be of use in materials testing. The development of new technology for isolation and propagation of potentially useful functional human and other animal cell lines is discussed and referenced. The advantages in the use of standardized cell lines for materials testing are listed and emphasized.

KEY WORDS: cultured cell lines, standardization, functional cells, cell bank, reference cultures.

Design of cell-culture methods and systems for the testing of materials for toxicity should include, wherever possible, standardized cell strains or lines. Use of such biologicals ensures that results of materials tests may be comparable both among different laboratories at any given period and within the same laboratory even over an extended time interval of 10 to 50 years or longer. Of course, to attain such comparability the test line(s) as well as other variables (sera, growth factors, regimen employed, and so forth) must each be normalized. While this fact may appear obvious, too often the potential for variation in cells and microorganisms undergoing continuous propagation or in strains obtained from differing sources is neglected or not recognized.

The American Type Culture Collection (ATCC) is a private nonprofit international resource dedicated to the collection, preservation, and distribution of well characterized cultures of living microorganisms and metazoan

[1] Head, Cell Culture Department, American Type Culture Collection, 12301 Parklawn Drive, Rockville, Md. 20852.

cells. Total culture holdings are now over 30 000, and distribution for 1982 exceeded 45 000. The Cell Culture Department, which is one of ten within the ATCC, currently maintains for distribution 565 lines from 45 different species. New programs that have been recently acquired will more than double these holdings in the next five years. Lines included now are used in toxicity testing, as substrates for viruses, in diverse basic research studies on cell physiology and growth, in cancer research, and for production of monoclonal antibodies to a variety of microorganisms and animal cells.

This report summarizes steps taken at the ATCC to collect or develop, characterize, bank and distribute cell lines for potential use in materials testing and for the wide range of other research and diagnostic applications previously indicated. Cell lines of possible special utility for materials testing will be briefly described.

Procedures

Cell Banking

Many of the methods used for cell banking and characterizations have been published elsewhere and will not be detailed here. Most investigators submit starter cultures without charge and with minimal restrictions on distribution. For some component lines recipients must sign a release form indicating that the cultures will not be redistributed or otherwise used for commercial purposes.

It should be emphasized that in those cases where investigators wish to apply for patents associated with given lines the procedures and forms required are different from those to be described below, at least until a patent is issued. Individuals or organizations interested specifically in patent deposit should contact the ATCC for pertinent information in that regard.

Required data for general submission of lines include a statement of exact species, strain, source tissue, and precise original and subsequent dissociation procedures plus indication of the growth medium and supplements chosen. A statement of the specific functional characteristic that would make appropriate strains of unique value is also requested. In addition, preference is indicated for inclusion of detailed histories on each strain; information concerning date of origin, sex, race, and age of donor; isolation procedures; inoculation densities used; and number of passages undergone as well as indication of the average number of cell generations accrued. Information on karyological analyses and survival potential of the cell lines is also solicited. Detailed descriptive data of the type indicated above is needed especially for inclusion of lines in the certified cell line (CCL) category. This grouping consists of populations that are the most rigidly characterized. In the absence of external funding, cell lines submitted to the ATCC are generally accessioned in the cell repository line (CRL) category. These holdings are tested for contaminating microbial organisms; their species of origin is verified, and recoverability from liquid nitrogen storage is documented. In general, no other characterizations are applied. Lines with very restricted doubling potential are also generally added to this CRL grouping.

After review of the credentials of each given line and consultation with advisors, decisions are made to proceed with initial characterizations. Generally starter cultures or ampules are obtained from the donor, and progeny are propagated according to instructions to yield the first "token" freeze. Procedures used for freezing cell lines have been described in detail elsewhere [*1*].[2] Complete information concerning the production of media at the ATCC and testing procedures for serum supplements have also been published [*2,3*]. Note that the quality control in this case includes tests for both adventitious agents and for growth promoting properties of fluids used. Cultures derived from such token material are then subjected to critical characterizations, and if these tests suggest that further efforts are warranted the material is expanded to produce seed and distribution stocks. The distribution stock consists of ampules that are distributed on request to investigators. The reference seed stock, on the other hand, is retained to generate further distribution stocks as the initial stock becomes depleted.

These various steps in the overall accessioning scheme are summarized in Fig. 1. It is important to note that the characterized seed stock serves as a frozen "reservoir" for production of new distribution stocks over the years.

Advisory Committee and Review

The ATCC consistently relies on advice from ad hoc consultants and more formally organized committees. The degree and mode of interaction vary somewhat depending upon the department, discipline, and source of support

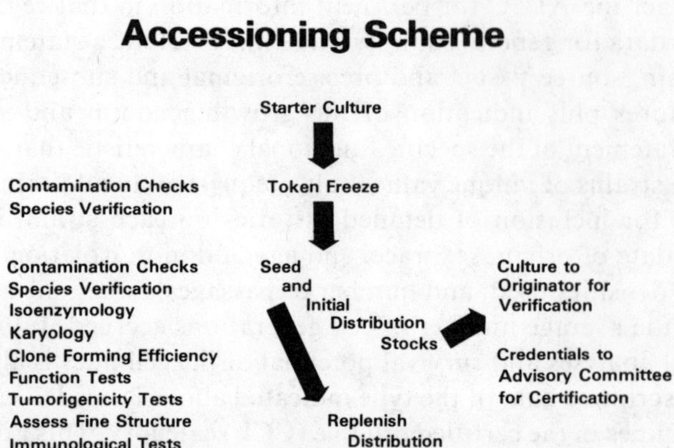

FIG. 1—*Scheme illustrating steps taken in accessioning cell lines and maintaining cell stocks at the ATCC.*

[2] The italic numbers in brackets refer to the list of references appended to this paper.

for differing accessioning programs. In certification of cell lines, for example, input is requested concerning lines to be added, characterizations to be applied, and descriptions on the cellular material eventually banked. Committee members are often consulted at points during the accessioning procedures and especially for the final certification process. In this case, complete descriptions and pertinent data are circulated for critical review. The cell lines and specific descriptions are considered acceptable if a majority of the committee members indicate concurrence.

A policy of returning both a typical distribution ampule or live culture plus the proposed catalog description and data description on each cell line to its respective donor has been adopted by the ATCC. The donating scientist is thus given the opportunity to examine material prepared at the ATCC to verify that the essential characteristics are retained. Descriptions on cell cultures are also returned to the donor for suggested additions or revisions.

Development of Cell Lines

Grant and contract-supported projects for the development of cell lines at the ATCC have been described elsewhere [*4–6*]. The ATCC Catalogue of Strains II, available without charge (except for international postage), includes many of the cell lines developed through such support. Established methods for accessioning and certifying this material are also applied for lines developed here.

Characterizations

Tests for Adventitious Agents—Routine tests applied at the ATCC to detect the presence of adventitious agents in cultured cell lines have been published. Thus Cour et al [*7*] described methods for detection of bacterial and fungal contaminants in cultured cell lines, and Dilworth et al [*8*] detailed techniques for detecting protozoa. Hay et al [*9*] presented techniques for detecting viruses in cell cultures, and Aldrich and Macik [*10*] provided a stepwise procedure for reverse transcriptase assay to be applied as an indicator of oncogenic ribonucleic acid (RNA) viruses in cell culture fluids. Two methods for detecting mycoplasma are used routinely at the ATCC. The first is a modified culture technique while the second relies on the Hoechst fluorescent stain (Fig. 2). Both methods have been outlined by Macy [*11*].

Tests for bacteria, fungi, protozoa, viruses, and mycoplasma are applied first to token freeze material and again during the expansion process. These are repeated on pooled material prepared from distribution and seed stocks, and the test for reverse transcriptase is also applied to culture fluids from cells reconstituted at this later stage.

Species Verification and Tests for Intraspecies Cell Cross Contamination—The problem of interspecies and intraspecies contamination among cell lines has been well documented [*12,13*]. At the ATCC cell line species is verified

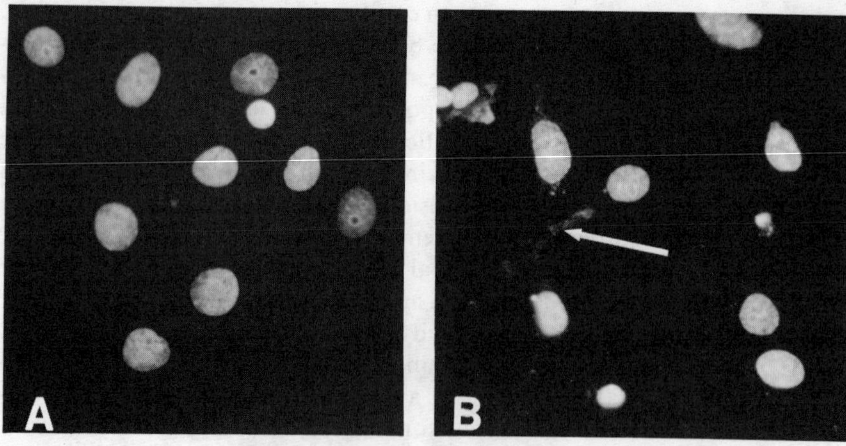

FIG. 2—*Hoechst stain to detect mycoplasma contamination in cultured cells:* (A) *control cell line without contamination and* (B) *contaminated cell line. Arrow indicates mycoplasma contaminants. See Ref* 11 *for methodology.*

both by staining with a species-specific fluorescent antibody [*14*] (Fig. 3) and by determining the isoenzyme profiles of glucose-6-phosphate dehydrogenase, lactic dehydrogenase, and nucleoside phosphorylase using starch gel electrophoresis [*15*]. These are sufficient to verify species of origin.

Intraspecies contamination of human cell lines may be detected by defining the isoenzyme profiles of a group of enzymes as described by O'Brien et al [*16*]. Alternatively or in addition, human blood group antigens and human leukocyte antigens may be detected using suitable reagents [*17,18*].

Karyology—Karyological analyses are performed on all cell lines included

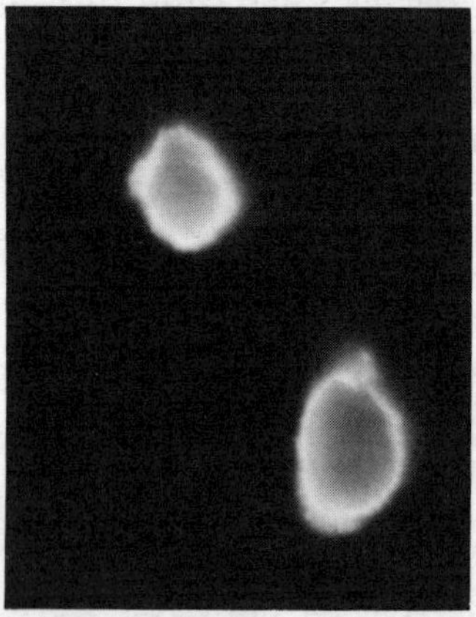

FIG. 3—*Fluorescent antibody stain to identify cell species. The halo around these two cells is typical of a positive reaction. Reference* 14 *gives methodology.*

in the certified category. Chromosome preparations from monolayer cultures are stained with aceto-carmine (conventional staining) or by G-banding [*19*] and Q-banding [*20*] procedures. Fifty metaphases from slides stained with aceto-carmine are examined for determination of the modal chromosome number and structural chromosome abnormalities. Karyotype constructions are made and compared with those published in the literature for confirmation of species. G-banded karyotypes are analyzed in selected cell lines to detect marker chromosomes (Fig. 4). Q-banding is done on all human cell lines to determine the sexual origin (presence or absence of the Y chromosome).

Plating Efficiences and Documentation of Cell Recovery from the Frozen State—A cell suspension pooled from the ampules of the first seed and distribution stock is serially diluted through ten steps, and aliquots are seeded directly to T-75 culture flasks in the recommended growth medium. These are then incubated for a period of ten to twelve days. The resulting cultures of cell colonies are fixed in buffered glutaraldehyde and are stained with a 1% aqueous solution of toluidene blue. Clones (Fig. 5) consisting of 16 or more cells are then scored to provide an estimate of plating efficiency that is generally recorded in the ATCC Catalogue of Strains II. Note that the inoculum for these tests comes directly from the frozen stocks not from logarithmically growing cell populations. This has been the standard procedure for all certified lines at the ATCC.

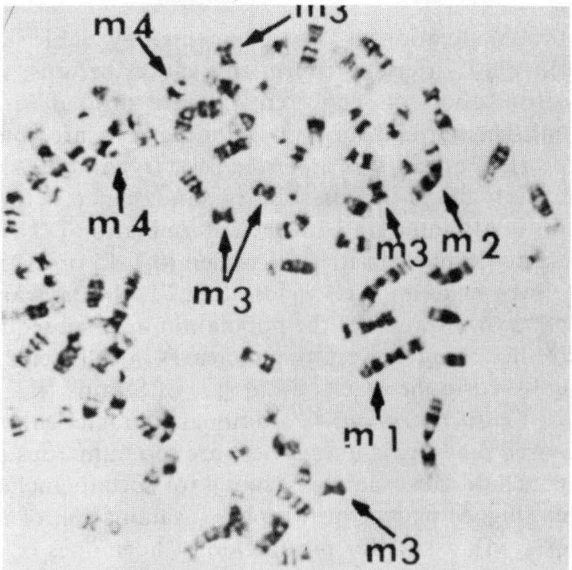

FIG. 4—*Giemsa banded karyotype of the cell line ATCC.CCL 2 (HeLa). Marker chromosomes* (m1 to m4) *are indicated by the appropriately numbered arrows. The preparation was as described by Lavappa* [19].

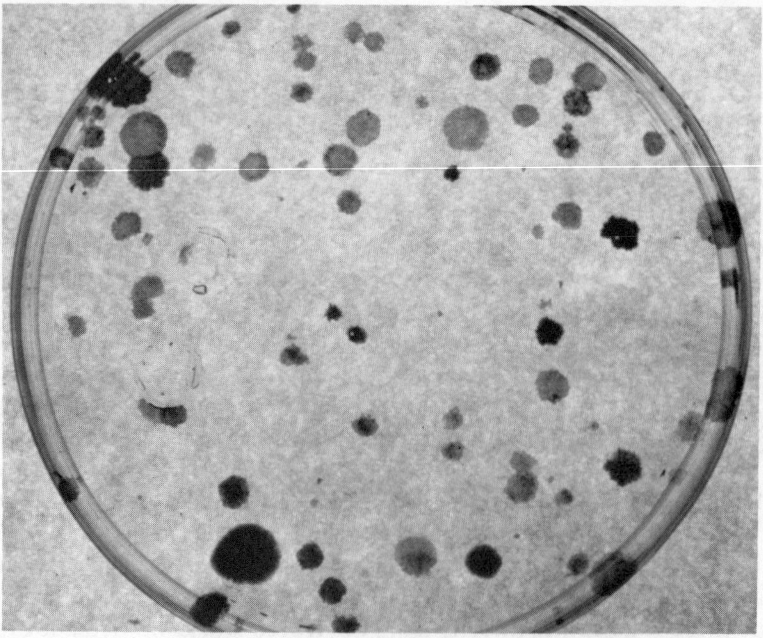

FIG. 5—*The quality of serum lots, culture media, or recoverability of frozen cell stocks may be quantitated by clonal analysis as shown here. Single cells proliferate after adhering to form large colonies on the culture plate* [1,2].

Recovery of cell populations is also documented by viable staining and by determining cell yields after short term cultivation periods. In the former test, aliquots from pools of recovered cells are diluted in erythrocin B (100 mg/100 mL saline), and the dye-excluding cells are counted using a hemocytometer. (Erythrocin B is preferred over trypan blue since the latter has a greater affinity for protein in solution, and, hence, serum or released cellular proteins could interfere with the determination [*21*].) In the latter case cell inocula at various densities are seeded to T-15 or T-25 flasks in the appropriate growth medium. Cell yields after 7 to 14 days are then determined by simple trypsinization of the population with subsequent quantitation using a Coulter counter. Resulting increases in cell yields are again recorded and published in the ATCC Catalogue of Strains II.

Miscellaneous Characterizations—Additional characterizations of diverse nature are required on specific lines. These are too numerous to describe in detail but may include, for example, staining for specific inclusions such as lamellar bodies (Fig. 6) or mucous droplets; examination of ultrastructure (for desmosomes) [*6*]; assays for production of hormones or other factors (gonadotropin, prostaglandins, CEA, steroids, and so forth) or tests for type and specificity of immunoglobulins released or both. The longevity of lines that exhibit senescence can also be quantitated [*6*].

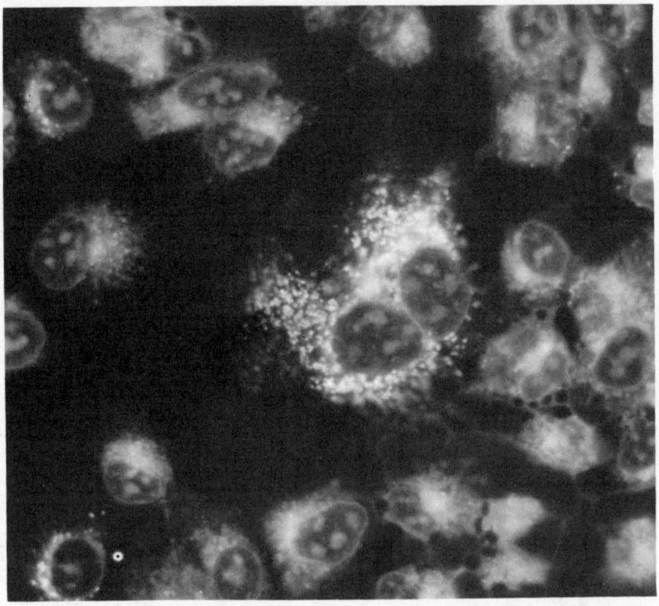

FIG. 6—*Some cell types may be characterized by staining for specific inclusions. This line (ATCC.CCL 185-A549) was treated with the fluorescent dye, phosphine 3R, to demonstrate characteristic phospholipid inclusions.*

Cataloguing—Detailed characteristics of lines in the certified category are listed in the ATCC Catalogue of Strains II that is published every two years. The most recent, the fourth edition, was released in Jan. 1983. Listings of lines accessioned during the biannual interim are included in a newsletter that is published every three months and is distributed to over 40 000 recipients.

Results

It is beyond the scope of this presentation to include comment on all potentially useful cell lines banked at the ATCC. However, Table 1 lists some of the available cell lines with characteristics that may be of special interest in toxicity testing. Lines that are simple to propagate and manipulate (for example, CCL 1 and 2) or have simple karyology (for example, CCL 16 and 29) offer advantages that have been recognized by many testing laboratories. Lines with sensitivity to contact inhibition (for example, CCL 92 or 163 [3T3]) that can be readily transformed by chemicals [*22*] (for example, CCL 226, $C_3H10T\frac{1}{2}$ [Fig. 7]) may deserve special emphasis.

Many other cell lines that exhibit the functional traits of their source organs (for example, CRL 1444, 1446, and 1458 [rat smooth, heart, and skeletal muscle lines]; CCL 207 and 209 [bovine endothelial lines]; and CRL 1468

TABLE 1—*Example cell lines of potential use in materials testing.*[a]

ATCC Number	Designation	Donor	Species and Tissue of Origin	Pertinent Characteristics or Comments or Both
CCL 1	NCTC clone 929(L)	Earle	mouse connective tissue	simple propagation
CCL 2	HeLa	Scherer	human cervical carcinoma	simple propagation
CCL 16/39	Don/Dede	Hsu	Chinese hamster lung	simple karyotype and propagation
CCL 34	MDCK	Madin/Darby	dog kidney	produces hemicysts or "domes" in culture
CCL 46.1	NCTC 3749	Evans	mouse lymphoid neoplasm	produces interleukin 1
CCL 75	WI-38	Hayflick	human lung	standard diploid strain
CCL 79	Y-1	Sato	mouse adrenal tumor	produces steroid hormones
CCL 82	GH_1	Sato	rat pituitary tumor	secretes somatotrophin
CCL 92/163	3T3	Green/Aaronson	mouse embryos	sensitive to contact inhibition
CCL 98	BeWo	Pattillo	human choriocarcinoma	produces gonadotropin lactogen and steroids
CCL 106	$LLC\text{-}RK_1$	Hull	rabbit kidney	epithelial-like
CCL 156	RPMI 1788	Moore	human peripheral blood	from an apparently normal individual; secretes immunoglobulin *M* lambda chains

CCL 207/209	CPA/CPAE	Ryan/Del Vecchio	cow pulmonary artery	endothelial lines, positive for angiotensin converting enzyme
CCL 208	4MBr-5	Caputo	monkey bronchus	epithelial line, PAS positive inclusions
CCL 218	WiDr	Noguchi	human colon tumor	produces CEA, tumorigenic in mice
CCL 226	$C_3H10T\frac{1}{2}$	Heidelberger	mouse embryo	contact sensitive; easily transformed
CCL 239	HCMC	Danes	human colon mucosa	epithelial-like, diploid
CRL 1442	BRL3A	Nissley	rat liver	epithelial-like, produces MSA
CRL 1444	A7r5	Carlisle	rat aorta	smooth muscle
CRL 1446	H9c2(2-1)	Carlisle	rat heart	forms multinucleated myotubes
CRL 1448	L6	Schubert	rat thigh muscle	forms myotubes
CRL 1468	FRTL	Coon	rat thyroid	diploid and epithelial-like, produces thyroglobulin not tumorigenic
CRL 1571	NRK-52K	De Larco	rat kidney	epithelial-like, transformation properties described
CRL 1600	H4-II-E-C3	Becker	rat hepatoma	epithelial-like, synthesizes serum proteins and responds to hormones

[a] See ATCC Catalogue of Strains II for more information or contact ATCC at 800-638-6597.

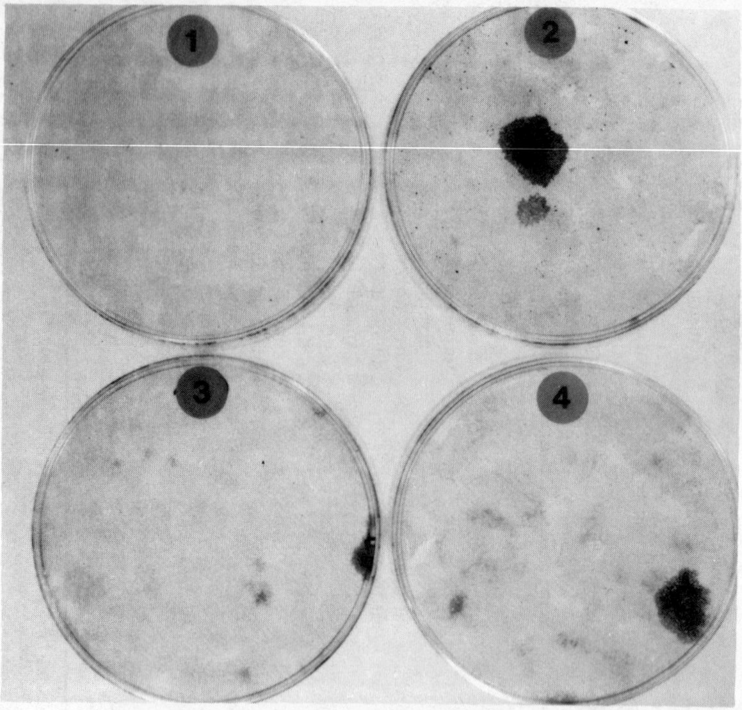

FIG. 7—*Cells of ATCC.CCL 226 ($C_3H10T\frac{1}{2}$) can be used to test for oncogenic transformation. Plate* 1 *is a control showing a contact inhibited monolayer. Plates* 2, 3, *and* 4 *contain colonies or foci of transformed cells that arose after cells were exposed briefly to the carcinogen methyl cholanthrene* [22].

[rat thyroid line]) have received comparatively little attention in this regard. However, they may offer special utility in tests for effects of materials on the in vivo function of cells from these organs.

Discussion

The procedures described for addition of human and other animal cell lines to the ATCC provide standardized stocks for research and industrial uses. The availability of characterized reference cultures stored in liquid nitrogen provides the following significant advantages to investigators and commercial organizations:

(1) the investigator can obtain contaminant-free standard cells at the same passage level from time to time over the years for systematic experimentation;

(2) cells whose origin, history, and characteristics have been described and documented are thus readily available;

(3) a source of types of cells that may normally be difficult to acquire is described;

(4) the responsible investigator is released from time-consuming and costly long-term serial subcultivation of cells; and

(5) both individuals and institutions have insurance against the total loss of valuable cells through contamination, alteration, or accident.

The advantages of working with properly identified cells in a system free from contaminating organisms is obvious enough in terms of conservation of time, effort, and research dollars. Of additional importance are advantages to be gained in the future by cooperative and coordinated efforts to characterize cells more completely as well as efforts to improve and standardize procedures and culture media.

An additional overall advantage to the entire scientific community is the provision of a common source of certified cells at similar passage levels that will be available for repeated reference indefinitely. It is well known that the characteristics and properties (proliferation rates, structure of chromosomes, virus susceptibilities, and so forth) sometimes change markedly when animal cells are cultivated for extended periods in vitro. Thus cell populations developed and maintained in different laboratories may vary widely. The availability of a common source of cells through the ATCC increases the possibility that meaningful comparative studies can be made both within a given laboratory and among different laboratories throughout the world. The ultimate advantages of working with characterized cells "captured" at a specific stage in their life history cannot be overemphasized.

The recent development of methodology for isolation and propagation of functional cell lines, especially from human sources, may provide additional cell lines of utility for materials testing. Thus, for example, human epidermal keratinocytes [*23,24*] not yet banked at the ATCC may be extremely useful in this area. Similarly, human bronchial epithelia may now be propagated in sufficient quantity for banking, standardization, and distribution [*25*]. Advances in our understanding of cell requirements for organic matrices [*26*] and humoral growth factors [*27*] may well permit like development for banking and utilization of functional cell lines from many other human tissues in the not too distant future.

Acknowledgment

This work is supported under contract N01-RR-1-2101 from the National Institutes of Health, Division of Research Resources.

References

[1] Hay, R. J., *Tissue Culture Association (TCA) Manual,* Vol. 4, Gaithersburg, Md., 1978, pp. 787–790.

[2] Caron, M. P. and Hay, R. J. *TCA Manual,* Vol. 5, 1979, pp. 1111–1115.

[3] Kern, J. and Franke, J., *TCA Manual,* Vol. 5, 1979, pp. 1131–1133.

[4] Hay, R. J. Cell Populations, Methodology Surveys (B): Biochemistry, Vol. 8, E. Reid, Ed., Wiley, New York, 1979, pp. 143–160.

[5] Ruoff, N. and Hay, R. J., *Cell and Tissue Research,* Vol. 204, 1979, pp. 243–252.

[6] Hay, R. J., Williams, C. D., Macy, M. L., and Lavappa, K. S., *American Review of Respiratory Diseases,* Vol. 125, 1982, pp. 222–232.

[7] Cour, I., Maxwell, G., and Hay, R. J. *TCA Manual,* Vol. 5, 1979, pp. 1157–1160.

[8] Dilworth, S., Hay, R. J., and Daggett, P.-M., *TCA Manual,* Vol. 5, 1979, pp. 1107–1110.

[9] Hay, R. J., Kern, J., and Caputo, J., *TCA Manual,* Vol. 5, 1979, pp. 1127–1130.

[10] Aldrich, C. D. and Macik, P., *TCA Manual,* Vol. 5, 1979, pp. 1147–1150.

[11] Macy, M. L., *TCA Manual,* Vol. 5, 1979, pp. 1151–1156.

[12] Gartler, S. M., *Nature,* Vol. 217, 1968, pp. 750–751.

[13] Nelson-Rees, W., Daniels, W. W., and Flandermeyer, R. R., *Science,* Vol. 212, No. 4493, April 1981, pp. 446–452.

[14] Stulberg, C. S., Simpson, W. F., and Berman, L., *Proceedings of the Society of Experimental Biology and Medicine,* Vol. 108, No. 2, Nov. 1961, pp. 434–439.

[15] Macy, M. L., *TCA Manual,* Vol. 4, 1978, pp. 833–836.

[16] O'Brien, S. J., Kleiner, G., Alson, R., and Shannon, J. E., *Science,* Vol. 195, No. 4284, March 1977, pp. 1345–1348.

[17] Kornbluth, J., Pollack, M. S., Fogh, J., Carey, T., and Dupont, B., *Transplantation Proceedings,* Vol. 10, 1978, pp. 735–738.

[18] Espmark, A., Ahlgrist-Roth, L., Sarne, L., and Persson, A., *Tissue Antigens,* Vol. 11, 1978, pp. 279–286.

[19] Lavappa, K. S., *TCA Manual*, Vol. 4, 1978, pp. 761–763.

[20] Caspersson, T., Zech. L., Modest, E. J., Foley, G. E., Wagh, U., and Simonsson, E., *Experimental Cell Research*, Vol. 58, 1969, pp. 141–152.

[21] Phillips, H. J., *Tissue Culture: Methods and Applications*, P. F. Kruse and M. K. Patterson, Jr., Eds., Academic Press, New York, 1973, pp. 406–408.

[22] Reznifoff, C. A., Bertram, H. S., Brankow, D. W., and Heidelberger, C., *Cancer Research*, Vol. 33, Dec. 1973, pp. 3239–3249.

[23] Rheinwald, J. G. and Green, H., *Nature,* Vol. 265, No. 5593, Feb. 1977, pp. 421–423.

[24] Tsao, M. C., Walthall, B. J., and Ham, R. G., *Journal Cell Physiology,* Vol. 110, 1982, pp. 219–229.

[25] Lechner, J., Haugen, A., Autrup, H., McClendon, I., Trump, B. F., and Harris, C., *Cancer Research,* Vol. 41, No. 6, 1981, pp. 2294–2304.

[26] Kleinman, H. K., Klebe, R. J., and Martin, G. R., *Journal of Cell Biology,* Vol. 88, 1981, pp. 473–486.

[27] Barnes, D. and Sato, G., *Analytical Biochemistry,* Vol. 102, 1980, pp. 255–270.

Stanley A. Brown, Thomas D. Nickel, and Thomas W. Walker

Workshop on Cell-Culture Test Methods

This workshop was held as a continuation of the symposium. The purpose of the workshop was to discuss the ideas expressed in the individual presentations, and to discuss issues that would provide guidance to the biocompatibility section of ASTM Committee F-4 on Medical and Surgical Materials and Devices. The workshop was divided into four parts; in the first three, the discussion was focused on issues raised by the chairman, in the forth, written questions from the morning session were read and discussed.

Part 1: Cell-Culture Standardization

S. Brown (*chairman of symposium, University of California, Davis*)—I would like to welcome you all to this workshop on cell-culture testing of medical and surgical materials and devices. We have heard a number of excellent papers in which the authors discussed their methods and results, and the rationales for selecting their particular tests. We would now like to discuss the ideas expressed in these papers, with the goal being the development of standardized methods for cytotoxicity testing. I would first like to direct the discussion to address several issues that have been the topic of many task force and biocompatibility section meetings. We will then read the written questions that you have submitted and have them answered by the individual speakers and discussed openly.

The first question I would like the discussion to address is, in essence, what should the biocompatibility section be doing about cytotoxicity cell-culture test methods? Should we write standards? Should they be generic or specific? If we write them, would anyone read and use them?

R. Hay (*American Type Culture Collection*)—I would like to ask a question out of ignorance, which relates to, but does not answer your question. Are there governmental regulations on materials that are being produced by the manufacturers in this group that impose some kind of control?

T. Nickel (*Travenol*)—Yes and No. There are no specific regulations on material requiring tissue-culture testing. However, many products are required to be cleared through the Food and Drug Administration (FDA) and other agencies. Scientific submissions are required, and tissue-culture testing would be included.

R. Hay—I asked the question because it would be better for a group such as ASTM to come up with the regulations, recommendations, or guidelines, in answer to your question, rather than have the government step in and do that for you. It would seem to me that it would be very beneficial to have ASTM come out with guidelines.

T. Nickel—I think one of the problems with that is "is it doable?" We have been wrestling with that for some time, and to date it has not been. I think we have learned an awful lot today, and hopefully some new light will be shed on this. If we do write some standards, would anyone follow them?

K. Sumner (*Zimmer*)—In answer to your question, I think it is very important that some organization, and this one might be as good as any, put down some procedures that can be followed by a large body of testing people, whether it is private industry or contract laboratories. Right now there is the United States Pharmacopeia (USP), which is very vague and not at all specific as to how it is done, what controls should be used, and so on. I think it is very evident from what we have heard today that there are a variety of people who can contribute towards specifying how these tests should be done. It would be very helpful to manufacturers to specify that a testing organization follow ASTM standards.

E. Frisch (*Dow Corning*)—I think the goal is to develop a standard that can become a voluntary or regulatory standard as part of the biocompatibility tactics that people offer with their products. I would like to address a question, also, as to whether manufacturers follow these? I think it depends a lot on the standard. I know that in our own case we have had about 15 years of experience using a tissue-cell culture method for silicone implants. I think it is safe to say that we have made a larger number of silicone implants than any other manufacturer. The materials that have gone through this test program have been used by many manufacturers. Millions of implants have been produced and quality assured by this method. It seems to have a proven history of sensitivity, specifically for silicone elastomers. We base all of our continued historic clinical and biocompatibility data on the fact that this test is reliable. So when we test material by this cell-culture method, we can be assured that our two- and three-year animal implant data apply. It would be very hazardous for us to change it at this point because we could lose all of the long-term testing that we have done and all of the extensive clinical experience that has been gained on it. We probably would not change unless this method were incorporated into the total test method package.

J. Black (*University of Pennsylvania*)—Having worked, as it were, from the beginning helping to initiate some of these considerations within ASTM, I would like to make two comments. First of all, I think the time is clearly ripe for at least two consensus test methods. From what I have heard this morning it would be possible to develop a consensus test method for fibroblast cytotoxicity in tissue culture, and it is probably also possible but somewhat more difficult to develop a test method for fibroblast growth inhibition in tissue culture. Now, I expect the hardest part of writing both of those standards would be the scope and rationale statements, because there seems to be a considerable difference of opinion on the applicability of these test methods. I use the terms "growth inhibition and cytotoxicity" very specifically, because I do not believe either of these test methods are even in a narrow sense biocompatibility test methods. They are simply as stated: one for cytotoxicity and one for growth inhibition. To the second question, "Would any one use them?" the answer is yes and no. I suspect they would be very useful test methods, frequently referenced and would serve very much the same role that ASTM Practice for Assessment of Compatibility of Metallic Materials for Surgical Implants with Respect to Effect of Materials on Tissue (F 361) and ASTM Practice for Assessment Compatibility of Nonporous Polymeric Materials for Surgical Implants with Regard to Effect of Materials on Tissue (F 469) currently serve, that is, as benchmarks, measures of what consensus is 1982, 1983, or 1984 when they are adopted, against which people will measure their test methods. I think that is very important because the previous speaker mentioned the major methods in use today are the USP methods, and they are so fuzzy that it is very difficult to make a measurement against them. It is very clear, in fact, from my discussion with many manufacturers, that most manufacturers who say they use USP test methods, in fact, do not. Because they have made so many variations and substitutions in the method as Eldon Frisch suggested, for their own use and needs, that the original USP methods can no longer be considered accurate benchmarks. I think that the time is probably ripe in these two areas for new benchmarks to be established.

S. Brown—Basically, then, what we might be developing are primary standards, against which one would compare a specific test. This would not be done with the expectation that a particular lab is going to change, but at least they could say that their results are consistent with these standardized methods. Thus, we would have some common denominator, which is, of course, one of the purposes of standards, so that we can all talk within the same frame of reference.

J. Black—Absolutely.

H. Shaeffer (*Finley Research*)—With regard to the use of such test methods, I work at a contract testing laboratory and deal with a lot of small customers, as many device manufacturers do. These people, I think, would like to see a test more legitimized by having an organization such as this write it down and make it official. Unlike large organizations such as Dow Corning who feel a little bit more comfortable about going out on their own, these small manufacturers like to say, "This is a USP test or ASTM test." In this way, an ASTM method would be useful.

S. Brown—Yes, I sensed that this morning, in a couple of discussions about contract biological testing labs where a customer would contact the contract lab, and the lab would recommend a couple of possible test methods. We could provide a service there by having some standards that they would feel more comfortable with.

J. Lemons (*University of Alabama*)—I would like to look back about five years or more to when we started talking about this. The question we addressed related to the specifics of how we would run the test. We became involved in this and found that depending upon those items that we purchased and the way we ran our test, we obtained quite different results. I would like to actively support the statements made here for a need for standardized testing. But I think we are going to have to be very cautious to specify each item very carefully because depending upon the glassware you use and the dishes you use for containing the cells, and the media that you use, you can really change the results quite significantly. We got out of the business entirely a very long time ago, because we found it difficult to understand all of these variables. I do not mean to throw it into a research era, but I mean to suggest that if you were to have a set of knowns, which we could work towards or at least understand, it would provide a comparable base. But I really think that this document needs to be very specific in terms of each item that would be used. That is a recommendation.

S. Brown—That has been the debate and controversy for the last five years: should we write a generic, all-encompassing kind of document, or should we write one that is specific as a cookbook that would be a solid benchmark.

J. Lemons—I am afraid that you would just confuse the literature more if you focused the distribution of information that exists here in terms of the variance that exists in test methodology.

L. Veresh (*Dow Corning*)—One gentleman was commenting that it would be helpful if there were a couple of tests that were sort of standard, which people could refer to. He was saying that for small manufacturers, but I think that is not only true for small manufacturers, but others as well. Perhaps people in development that do not really understand some of the more esoteric parts of tissue culture and would give them something that they could kind of hang their hats on that would not require a lot of detailed experience with tissue culture or knowledge of tissue culture that they could talk about, work with, and use in evaluating materials.

J. Lontz (*Veterans Administration Medical Center*)—I think first that the idea of a voluntary agency, like ASTM, proposing a test provides a broad range of representation such as present here. There are other agencies such as the Society for Biomaterials and professional medical societies, especially concerned with plastic and reconstructive surgery. But I do not see them moving any faster than we in ASTM would, nor providing a broad consensus coverage such as representation from government, academia, small and large contractors (producers) compared to select medical and scientific agencies. Second, I think the cell-culture test methods should be highly objective rather than purely empirical and archaic tests (that never die) with archaic antecedence that will hardly match up with the rapid progress now evident in tissue-culture science. (Note this is a modification and pointed simplification of the original oral comment.) At the same time, if need be for objectivity and specificity there is no reason why there cannot be two or three good cell-culture biocompatibility tests, considering that biomaterials serve so many different medical and surgical uses. Let us have two or three good tests, why not?

R. Wallin (North American Science Association)—Dr. Brown, to get back to your question, "Is it appropriate for ASTM to develop this standard and would anyone use it?" I am curious about your question. ASTM has been involved in writing that standard as well as other biocompatibility standards for some time.

S. Brown—Yes, but the effort represents about six years of frustration, during which time we continued to ask questions about what type of test we should write, and would anyone use it.

R. Wallin—Is that not true of the standards for other biocompatibility tests too?

S. Brown—Yes. My feeling, actually, is that we have come very close to a meeting of the minds and I am quite optimistic that we can make progress. The thing we want to keep asking ourselves as we move forward in this business is, Are we simply flapping our wings and writing documents that nobody will ever read? I have heard a number of references at this meeting to the USP surface area for extraction, yet ASTM Committee F-4 has had an extraction method ASTM Practice for Extraction of Medical Plastics (F 619) for three years. I do not know if it means no one has read our extraction protocol. We have also heard reference to the USP short-term implant and systemic toxicity test methods. ASTM Committee F-4 has just passed similar documents ASTM Practice for Short-Term Screening of Implant Materials (F 763) and Practice for Evaluating Material Extracts by Systemic Injection in the Mouse (F 750), and it will be interesting to see if anyone references them. I hope they will. We just want to make sure as we embark on what may be a mammoth task, there is a need. I sense from this discussion that we should regroup and continue to move forward. We actually have one document that, with the exception of a couple of negatives, represents the consensus. The negatives are such things as cell line. I think this benchmark philosophy will enable us to move the document through to society ballot.

J. Black—I echo your comments about frustration, and I was suggesting in my earlier comments that the most difficult parts of the standards to write will be the scope and rationale. When you get those right and get consensus on that, everything else follows as night follows day.

S. Brown—And that is the key: what is the application of the test, when is it to be used, or is it simply a reference against which other test methods can be compared? As a benchmark, we could perhaps get something moving forward.

We started out with a question of whether ASTM should develop cell-culture test methods. I sense a strong consensus of agreement. Rather than taking a poll, let me put it another way, is there anyone in here who would strongly recommend that we not pursue the development of some benchmark cell-culture test methods?

No response.

Well, then I guess this meeting was worthwhile. We have accomplished one major objective and that is to get a vote of confidence to continue. There are 5 h of meetings scheduled over the next three days. I hope you will all join us and continue working.

Part 2: Standard Charts or Methods for Ranking Cell Reactions

S. Brown—Another specific question I would like to field is a comment I made this morning. Should we pursue the development of standard reference charts? Sort of using the analogy of ASTM standard for Estimating the Average Grain Size of Metals (E 112), reference charts for metal grain size and inclusion content are just wall hangings of metallographic sections that are all taken at the same magnification. To use them, you sit at a microscope looking at a piece of metal, you look at the wall, and you can just read off the number in terms of, in that case, the size of grains. I think from some of what we have seen this morning we could do the same kind of thing with some fairly standard grading schemes from 1 to 5, and use that as another way of coming to grips with the issue of standardization. We could develop charts for different cells and maybe different test preparations.

Comment—It would only work if you standardized the cells and procedures, because there is enough variation seen there with different methodology and cells. You usually look at L-929 and WI-38. I think the other problem is that no one has said where the cutoff point is: I have heard people say we ignore "1's," and others say we look at "1's."

S. Brown—This would be done on the assumption that they were developed in conjunction with a standardized method. But at least we could define what a "1" is. As a next step, we can address the question of where we draw the cutoff.

L. Veresh—Another comment is that you may be able to show pictures of what under certain conditions are normal looking cells and under those same conditions what are not normal looking cells. But I think you will have a difficult time trying to make gradations of 1 to 5, to get pictures of what will look representative to all the people in this room.

M. Brande (*Kendell*)—I have noticed in a lot of test materials with L-929s, although I have used WI-38s, even with both lines the cells show up differently with different materials under toxic conditions. And that, I think, is possibly going to complicate a chart.

J. Fehn (*Medtronic*)—I think the idea of having benchmark procedures is better than to say we are standardizing. I think if we say that it would be better to have benchmark procedures rather than standardized tests, which would be guidelines for people to use, it would be pretty hard to have pictures depicting what it would be (for example, 1, 2, 3, 4) when you have taken these benchmark tests and adapted them to your given situation.

J. Ulreich (*University of Arizona*)—The introduction of radioactive labeling completely eliminates human variability. Nobody is looking at cells and judging them. Second, I am not certain whether any of these are now available, but would there be any sort of optical scanning device that could scan a petri dish or a flask and read the cell surface area. Any time you have an instrument doing something instead of a human, it seems to be a lot less susceptible to error. If the instrument was calibrated to cell size, would that be an index of proliferation?

S. Brown—I do not know, that is a good point.

Comment—That would restrict the use and cost of the test too much.

R. Schlesinger (*Bio-Technics*)—I wonder if the purpose of the standard is to teach people all of the technology involved, or is it to guide people in the performance of the test? I think the preamble of the test should state that only individuals who have some knowledge about culture techniques should perform the test. Perhaps what we are trying to do is wrong. The standard is not supposed to lead people into the light from the darkness but to give somebody, as you say it, a benchmark. I think most of our standards are written that way, and if you do not have a certain degree of knowledge at your fingertips, you are not going to be able to perform the test.

S. Brown—I agree that we do not necessarily want to start at the beginning.

R. Schlesinger—You may limit the procedure if you are going to wait for a set of pictures that everybody will agree upon. We could spend many years trying to get uniformity of these pictures, and as you know, from cell line to cell line, from toxin to toxin, the effects on the cells vary. That is the reason that our laboratory provides photographs in our reports to document the type of toxic changes that occur. We should not limit ourselves to one specific positive and one specific negative control, but leave the option for different laboratories to select what works best in their specific applications.

S. Meyers (*Abbott*)—It seems to me we are trying to get out of some way of ranking the cell reaction without necessarily saying this is a pass or this is a failure. You have a negative control material or a positive control material, and you can observe your cell morphology in the presence of your positive or negative control material, and that gives you the upper and lower bounds of what you are looking for. After that, it may well depend from lab to lab, but it seems to me that it would be reasonably easy at that point to develop your own ranking system given those two finite bounds.

S. Brown—But then you get that gray area of the intermediate response.

S. Meyers—But nevertheless, that intermediate response may depend on each laboratory; it may depend upon the kinds of materials you are looking at. You might want a 0 to 2 as your total bounds, some people might like it to look like we use 0 to 4, but, nevertheless, you find your upper and your lower bounds. I think that would be doable.

Part 3: Standard Reference Materials

S. Brown—Well, I recognize that we perhaps should abandon the idea of the standard chart. But since you have started the subject, should we go on to the question of reference materials?

S. Meyers—I think a source of reference materials would be good. I heard latex mentioned as a positive control, and think I heard someone else mention polyvinyl chloride (PVC) with tin; as negative control we used polyethylene; polyurethanes were mentioned. It would be nice when you have a positive control to know how positive it is, or is it so toxic that nothing is going to cause it to survive, or is it a level of toxicity that if your cell line loses some of its sensitivity to toxic materials, it will pick it up?

T. Genova (*Ethicon*)—He brings up a good point regarding the use of positive controls. We really do need something that is going to allow you to monitor the day to day sensitivity of your cell line. You do not want something that is going to be a 5+ or a 4+; it will wipe out your cells everytime you use it. Thus, you are essentially getting no information back from that control material with regard to your own variability. It is important to have something with a 2+ or 3+ response so that you can monitor day-to-day variability of your test method.

S. Brown—I agree; you can add methanol and nothing will grow. That does not say much. Do you have a suggestion of a mildly or sensitively toxic material that can be standardized and made available?

T. Genova—A lot of work has been published with the tin compounds and perhaps that is an area that can be explored further.

L. Ettorre (*Extracorporeal*)—I think the importance of the tissue-culture test method in talking about positive and negative controls is, Point 1, we use both positive and negative controls. The second point I would like to make is there are a lot of companies here that have a long-term history (five to ten years), as Mr. Frisch mentioned, using their own specific positive and negative controls. Taking that into consideration, I think it is important that we do not specifically limit ourselves to one specific positive and one specific negative control. We should leave the variability and openness for different laboratories to pick and select specifically what works best for them in their various applications for their own products. Otherwise we are going to start the whole ball rolling again of groups stating, "My method is not available, therefore, I am not interested in supporting tissue-culture test method standards."

L. Stroempl (*North American Science Association, Inc.*)—I was just going to add a comment that the reason North American Science uses titration methods is because it is so difficult to find positive control materials that are not overwhelming in toxicity. That allows us to use materials that in their own right would be very toxic, but through the testing of serial dilutions we prepare from the initial extract, we are really able to get a lot of very comparative data about our cell-line sensitivity.

J. Black—I want to support that position very strongly. I think that trying to find standard materials is fraught with almost insurmountable difficulties, and I think it would make much more sense to select a single cytotoxic material such as phenol, which is widely available, and have the test method have some sort of internal titration system so as to establish an operating curve and find out where the 50% kill point or 50% inhibition point is. I think that is going to be a much more fruitful approach in the long run than to try and establish several sources of real materials.

R. Schlesinger—There is a precedent for that because ASTM specifically defines the properties of the positive control, but it does not specify positive control material. Perhaps some combination of using the titration technique, as we use, and a satisfactory control, which a laboratory has developed, could be provided for.

J. Lontz—In the matter of controls, specimens, positive (toxic) and negative (nontoxic), my comments relate to objecting to the use of "latex rubber, such as a specimen obtained from a Foley catheter" (Section 8.0 in ASTM Task Group F4.20.06.50) because of the extreme variability of the numerous proprietary compositions. Our own solvent extractions of about a score of such specimens of different proprietary sources have shown a variety of latex compounding additives, characterized by infrared scanning for significant chemical group frequencies. The proprietary compounding and processing additives include vulcanizing agents, accelerators, light and heat stabilizers, antioxidants, and so forth of variable identification and composition, as indicated also by source references in rubber technology. While it is commendable that current users of the cytotoxicity tests have some means for getting around this problem of latex (surgical grade presumably rubber specimens), other interested users and newcomers may not necessarily receive the equivalent latex rubber tubing, equivalent in imparting the negative reference control. Moreover, latex rubber tube source vendors may be changing the compounding and processing additives, for some cogent reason or product enhancement, that may not be the same five years hence. The user of the cytotoxicity should know what the ingredient additives are in the positive control, reference and chemically identified at specified levels. Unless the specimen is chemically identified as to what the ingredients are, I would not propose its use in the cytotoxicity test.

D. Magoffin (*Bio-Technics*)—Since the consensus seems to be that the testing method be a benchmark document, perhaps it would be worthwhile to establish a number of different control materials, both positive and negative, that are very carefully defined. A reference data base could be assembled based on the experience of the people here with certain control materials. Perhaps we could assemble a consensus on how each material performs and what dosage will give various graded toxic responses.

S. Brown—Ending up, in essence, with a round robin testing of some of the materials with a given test method.

D. Magoffin—Yes. Then you could refer in the document to the specific toxic response a given control material should exhibit when the assay is conducted according to the standard method.

R. Wallin—Would it help focus on what our purpose is by asking, "what is the problem we are trying to solve?" What problem is there about cell-culture methods in use that we can solve by developing a standard?

S. Meyers—You want to know how sensitive your cell culture is to a toxicity. In other words, when I get what I call a maximum toxic reaction, and you get that in your cell line, is it the same level of toxicity, or is my cell line just more sensitive than your cell line? If we have the same reference material, we can tell.

R. Wallin—That gets back to the use of appropriate controls.

S. Northup (*Travenol*)—Having worked with this group for five years I have shared the same frustrations as Stan Brown. I would really like to get the group to take a smaller bite. I would like us to write up the agar overlay technique, but instead of saying "confluent monolayer" or "near confluent monolayer," I would just like to say monolayer and let each lab determine their own adjective for monolayer. And for positive controls also, I think, as an initial step, if we get just the consensus document, which is still going to allow the individual labs to interpret, we can reconvene after a period of time and settle as a group of experts on the definition or a more clear definition of what we mean when we talk about monolayers, or positive or negative controls in that respect. But as a start, we would have a document. The ASTM standards require a five-year review of this. It is a new opportunity for us to come together and share our experience with the methodology we have developed.

R. Schlesinger—I agree with Sharon Northup. The only way you will get a standard through, and I think it will be a worthwhile benchmark, is to loosen up on these areas where we have the conflicts: the cell lines and the controls. I think it is possible to define controls that are adequate, and to define cell lines that are of concensus, and put in words like "or equivalent," which will allow the individual to determine the equivalence. Of course, I think its the only way we will get through the next few years. Otherwise you will have to convince everybody to stop what they are doing and change to the standard method, which is almost impossible.

E. Horowitz (*Johns Hopkins University*)—The purpose of developing the standard is at least as twofold. One purpose is to develop a methodology that provides a field test method. To start out, you have to proceed in a more generalized way. The other purpose is to provide an ASTM standard that is essentially a referee document. It may not be useful in everyday practice, but it would be a strict methodology that could be referred to by parties in dispute, or by parties that want to clarify some area that is unclear. The moment you mention the term "or equivalent," you have opened up so many different variables that no two individuals are likely to conduct the test in the same manner. You need to be careful about what kind of methodology you are going to develop: is it to be a practical field test type of procedure that you use everyday, or is it going to be a standard that will be used essentially to resolve disputes or to set some limits that field test methods can refer to?

S. Brown—I think we have two slightly different things that are coming out, and this may be part of what has been confusing us over the last couple of years. We started out saying that we wanted something very specific: a cookbook, a benchmark. Then Dr. Northup made the point that there are some terms that we were having trouble defining, such as "confluent monolayer." That got us heading in the direction of a more general document. Now what Dr. Horowitz is driving at, again, is that we should make it more specific. I think, Sharon, what you are driving at, is that concerning some of the definitions of terms, we should say "establish a monolayer," but not really be so strict about what word we use. But that we should be specific as to the details of the method: what kind of dish, type of cell line, incubation temperatures; all should be in specific terms. We should not get hung up where there is confusion simply in definition of terms. Am I right that you were not trying to say that we should go back down to a sort of general nonspecific document? What Dr. Horowitz is saying is that we should have something that could be a "referee," a solid document against which comparison could be made.

E. Horowitz—In addition, you could use a more general, or generic method, where you have more latitude in your procedure.

R. Johnson (Medical Inc.)—I come from a small medical device manufacturer. For people who have not been in tissue culture for years or have been doing tissue culture with contract labs or the smaller manufacturers of which there are many medical device manufacturers that are not large, it would help them greatly in trying to set up their own labs to have some kind of recommended cell line, some kind of recommended negative and positive controls, at least some recommendations from ASTM so that they have something to start out and something to go on. Then, later on, if they find something that works better, they can adapt it to their own laboratories. At this point there is nothing that someone can use and say we have this kind of recommendation that these other manufacturers have tried, and this has worked for them and we know, at least, that there is some history behind this.

P. Brown (ASTM)—There are a lot of subcommittees that seem to come up against the same problem; in cardiovascular I have heard it said that there are maybe ten different methods out there for screening materials and each laboratory thinks theirs is best. I would suggest something along the lines of a standard guide to critical parameters in the development of tissue-culture testing. This guide would tell people, if you are going to develop a test, here are some of the critical parameters you must have some control over. Do not lay out the test yet as a specific test. Jonathan Black said that is basically what ASTM F 361 is already, and that nobody actually uses it. Everybody uses it as a guide to make their own test.

J. Black—Yes, but let me point out by way of defense, the only reason that worked is that ASTM F 361 is very specific. It is possible to take ASTM F 361, if you have reasonable skill in the art and run a very specific experiment, and then to pass it along to another laboratory to have essentially the same experiment run. So I think that you probably serve the general function best by being very specific. I would like to propose the test that we use in our laboratory for protocols. That is; they have to pass what we call the "falling safe test," if after you finish writing it and you walk out of the laboratory on your way home, and you are flattened by a falling safe, then somebody with a similar level of training and knowledge in the art, but without specific knowledge of the intent of the protocol, can do the test and get the same result. And we found that to be an extremely useful test for completeness.

L. Veresh—What I have been hearing, from these last comments especially, is going back to what we started talking about at first, that is, not developing one general standard but three or four or five standards, which we heard of this morning, and then to let a small manufacturer sit down and intelligently read over three, four or five methods and choose for themselves the one they feel would be more ethical. Give people the choice to use their own intelligence to decide which is more applicable.

S. Brown—But I think it ought to be something that when they have finished reading it, they know what to do, or at least a good idea of what to do.

L. Veresh—They would know how to do that particular method but give them a choice.

E. Frisch—Methods for devices used in the body of short periods of time, or which contact only external surfaces of the body need not be as demanding as with implants. I heard some discussion earlier of a matrix, something like what you are doing in the biocompatibility testing system, this would be essentially a matrix within a matrix.

S. Brown—To follow up Peter's (Brown) comment about cardiovascular, I think we are much closer to an agreement in terms of a fairly small number of tests whereas the thrombogenicity issue is still wide open, and I think they have a long way to go before they can even get all of them in the same room in the same lodge.

E. Frisch—But then you allow the report to be written so that they can specify the cell line used and the conditions of the test.

R. Wallin—I still have a basic problem with what our purpose is. There are guidelines written by the Health Industry Manufacturers Association, (HIMA) the Pharmaceutical Manufacturer's Association (PMA), and the British Standards Institution. They deal with how you test different materials according to the way those materials are incorporated into the devices. They reference, in most cases the technical literature, for example, the agar overlay procedure by Guess and Autian. These are technical reports that describe how one goes about doing a test. These could perhaps be written in greater detail with more "bells and whistles" attached to them, but there is a scientific literature that describes those methods. What is it that we need to do here?

S. Brown—I think what we need to do, and I cannot reference the specific ones, and I have not used any of these methods so I am totally neutral, or ignorant, but you have all said we are using a modification of this method or that method, which means someone has gone back to the original literature and then gone into a laboratory and had to do a little bit of refinement to make it work effectively. That is why we called this group together, because these are people who have gone back into the lab and who have made the necessary adjustments so that the methods do work. We need to incorporate whatever those little things are, those "bells and whistles," into a document that will save the new folks, the little company, the problem of trying to reinvent and go through and find all the bells and whistles. Not come up with new methods, but simply take the original articles and certainly reference them in our standard and write them up into a form that has matured since the paper was published.

E. Horowitz—It goes beyond just helping small companies. It is really working toward tests made in one laboratory being compared to tests made in another laboratory. The size of the company has nothing to do with it. What we are trying to do here is to provide a protocol either loose or strict, that would permit someone to get the same results that you obtained with your test and so that there is agreement rather than disagreement not only in scientific results but in a commercial sense as well. We have found in many other areas, that the way to do this is not to have a loose or poorly specified document, but rather to have a well defined and specific protocol document with respect to procedure and testing. Then when we conduct a round robin test in a number of laboratories, the results will agree within some specified parameters.

L. Stroempl—I think one of the difficulties we are having is that we seem to be talking about "the" standard, "the" document. I think there is a real difficulty in trying to write a single standard designed about a single method that is only going to be applicable to one family of materials.

S. Brown—We have three in draft stage now, so there is no effort, at least in terms of trying to develop only one. We realize that at least three or four are necessary; you need direct contact, overlay, elution, and human tissue. We would all be kidding ourselves, as you point out, to try to do it with one.

R. Schlesinger—Even those three methods should probably be flexible without specifying one source company for materials.

S. Brown—If we were going to develop a benchmark document that can be used for making comparisons of different methods and different laboratories it has got to be tight, like ASTM F 361, which is like ABCD, and then make a variation of that theme. But we have to write the scope, significance, and rational statements such that it is made very clear that we are not implying that this is "the" way to do direct contact cell culture, but that it is simply a reference method based on the original literature and based on some established data that can be used for making comparisons and whatever, but not, and this is what the protocol document helps us do, not implying that this is "the" only one.

R. Schlesinger—You may be unable to perform the procedure if you do not have that type of container specified. Tightening to that extent tends not to increase the value of the method.

J. Black—It is necessary. If you are going to make the method work, you are going to have to do that. And that is what the problem is; it has to be a reproducible method. ASTM F 361 took twelve years to write, and a lot of that time had to do with these sorts of problems. We have only been, what, six years on these?

P. Brown—It has been the experience of a lot of other ASTM committees that the heart of any test is the precision or reproducibility between labs. You can write the greatest standard in the world, but if you cannot go out and reproduce it, it is probably not worth publishing at ASTM. Now, the problem arises when you have multiple methods, and you do not want to force people into a corner on one method. I have seen this problem in the Forensic Sciences Committee. On that committee every method is a potential court room situation. They wrote no methods because, for example, one method on blood stains was going to put all the other methods on a lower level of acceptability. So they avoided that whole situation. It is a possibility that you may want to go out and do some field testing to determine if there is comparable precision among the methods that already exist, or to determine what is the precision. Even that kind of information may be more helpful than just writing a method.

S. Brown—We heard that data today, and there are methods that we may not necessarily want to endorse, however, we certainly could write a document. These have been put through round robin testing, and the data are available. The important thing is to not make it sound as if we are saying this is the only way to do a particular type of test. That is simply a matter of how we write the rationale statements.

E. Horowitz—Sometimes it is useful to look at the analogs in the physical sciences. For example if you look into the ASTM methods on viscosity you will find at least a dozen different methods for getting at essentially the same physical chemical parameter. These need not be just a single cell-culture method, we could have a full series of them depending on whether the company has that equipment or whether there are people there accustomed to using it. So keep that in mind.

The other point is also on the same subject of viscosity; you have to be careful about making small changes in the method. Some years ago there was international round robin on a relatively simple thing like measuring the viscosity of polystyrene. Specimens were carefully prepared by one of the companies and sent to about a dozen national laboratories around the world. The results that came back were so bad that a committee reported that the specimens had degraded, therefore, the results were inconsistent. We later tested that specimen and found that nothing happened to it. People took license with their procedure, and they made what they *thought* were small convenient but nonsignificant changes, and it destroyed the precision and accuracy of the results. I think the same kind of thing applies here.

Part 4: Questions from the Symposium Attendees

T. Walker—Now we will begin to get your questions answered.

Question to Anyone—Has the Beckman Instruments Microtox® system been evaluated by any of the organizations represented here?

No Response—It looks like none.

A Question to L. Smith (*Dow Corning*)—How are various cell lines qualified to demonstrate equivalence? What procedures have been used, and what is the extent of replication to substantiate the equivalence? Have any cell lines been found to be generally not equivalent or unacceptable based on reproducibility?

L. Smith—The last cell line that we approved as equivalent to the WI-38 was the HFS218 human foreskin cells that were qualified about two years ago. At that point they were evaluated in concurrent testing briefly to ensure their suitability. They were then included into our normal tissue-culture routine, so that every time we tested positive controls, negative controls, or production specimens against the WI-38s we also tested them against the foreskin cells. This was done for a period of about three months covering over 500 specimens. We have made the assumption that over three months' time, we get enough of the retests, strange reactions to positives, and failures, so that we can ensure that this cell line will be sensitive as the WI-38s. Over the three months time they always produced the exact same reactions as the WI-38s. So on the basis of this data we have qualified them as having an equivalent sensitivity to the WI-38s.

Question to E. Dillingham and R. Wallin from J. Westrum—It would seem that extraction media for solutions may be as important, if not more important, than cell line or test methodology. In general, and specifically, what alternative extraction media have been investigated in a comparative study? What kinds of relationships have been, or must be, established between extraction media and possible in vivo leachables to assure an accurate test?

E. Dillingham (*University of Tennessee*)—The primary screen that we have used over the years has involved just four extraction media and a high temperature, 121°C 1-h extraction protocol. The object is to determine the total leachability of the material. We have never found, except on rare occasions, that the extracts show toxicity where direct tests on the material were negative. In such instances, degradation of the material may account for the toxicity of the extract, and the extract test is repeated using a 37°C five-day extract to confirm the presence of a toxic leachable substance in the material. A negative retest is consistent with the conclusion that the toxicity was induced by the high-temperature extraction.

We have not investigated alternative extraction protocols with respect to the primary screen tests. We have found, however, that in the isolated heart perfusion test aqueous extracts perform best. Complex "physiological" solutions give erratic results. Significant changes occur in the "physiological" solution during extraction of a material, over and above the simple extraction of leachable substances from the test material. Results are more reproducible using the aqueous extract to prepare "physiological" solutions for the test.

I agree that a well defined protocol is required for materials testing. However, you must look realistically at the performance of the test to determine how detailed the protocol must be. Elaborate specification should be avoided unless demonstrated to be essential to reproducibility. The specification of standard reference materials is important as is the specification of the cell line. An immortal cell line is more important to long-term reproducibility of results than the question of human versus other cell lines. Since most in vitro tests only discriminate low, medium, and high response, there is a real question as to how elaborate the specification of such tests should be, particularly for acute toxicity assessment.

L. Stroempl—I thought I might address the same question from our perspective. Indeed, the conditions of extraction are an important factor in reproducibility, not only in this test, but also in any of the other tests in which extracts are evaluated, such as USP intracutaneous and systemic evaluations. That is one reason why it would be difficult to use a material such as phenol as a control material. We have a limited amount of data in which we have compared various extraction conditions, such as 120°C with 50 or 70°C conditions to determine whether or not those are truly different or equivalent and so on. That is a question that we are frequently asked, and I wonder if that was not the reason the question was asked today. However, we have only a limited amount of data on that subject, and I would not be able to address that at this meeting.

E. Dillingham—We have looked at autoclave (121°C) extracts and 37°C five-day extracts. If a toxic leachable substance is present in the test material, the results are generally comparable. There are two factors involved in the toxicity of an extract. One is the intrinsic toxicity of the leachable(s), and the other is the concentration of the leachable in the extract that is strongly influenced by the lipophilicity or solubility of the leachable in the extraction medium. Similar apparent toxicity will be observed with different extraction times if the time is long enough to reach the solubility limits of the leachable components. Since leachable components of biomaterials differ widely in their solubilities and rates of leaching, no single extraction protocol is optimum for all materials.

The advantage of the autoclave extract is that it is rapid and produces a sterile extract that can be tested directly in in vitro tests. Low-temperature extraction does not assure sterility unless the material is sterilized initially and extracted under aseptic conditions, a more elaborate procedure.

Question for J. Lontz from J. P. Harvey (AAOS)—Does the size of the test object make a difference in response? Does the size of the prosthesis, even weight, make a difference to the tissues in contact?

J. Lontz—Size and weight? We have not the slightest idea at the present time. We are still developing it. We first used a 1 by 1 cm to $^1/_{10}$ mm in thickness, and I think that weighs about 16 mg. We still have to generate that data. But we are delighted that you raised the question because we wanted to account for the mass balance through surface versus volume as was mentioned before, which we think is an indication of how rapidly suspect toxicants leach out from our prostheses.

E. Dillingham—We have looked at particle size as it influences the toxicity of autoclave extracts. Particle size has little influence on apparent toxicity of those extracts. Although a standard particle size or total surface area is desirable for extraction protocols, I see no satisfactory way of standardizing that parameter for the diversity of biomaterials and devices to be tested except for sheets and films where the USP protocol can be easily used. For materials other than sheets and films we use a standard weight of material for extraction.

J. Lontz—When we started developing our test, we wanted to see how the cells grow, spread, attach, and colonize as the criteria of physical cell-test surface interaction that could preempt toxic reaction. We had to consider the physical cell-test surface interaction as important as the chemical cell-toxicant solute interaction. We used Teflon®, known for its chemical nonreactivity, as a model for the study of the physical interaction. However, as a 50% crystalline material, it is substantially opaque at ½ mm thickness with about 80% loss in transparency. This meant that one needs to use a thinner specimen, down to $^1/_{10}$ mm. For a 1-cm^2 test piece, one could visualize that the leachable toxicant would provide too low a concentration to impose any cell-growth inhibition. Such being the case, the mass balance aspect, taking into account the surface thinness ($^1/_{10}$ mm) for transparency to see (and photograph) the cell growth and the specimen volume ($^1/_{10}$ cm^3) to provide the toxic level, can become crucial to the interpretation of the test on concentration of suspect toxicant. Put another way, if too dilute or too low in concentration the result could be a false negative. Incidentally, many of our orofacial cells do not grow over the specimen if the thickness is over about $^8/_{10}$ mm.

E. Dillingham—The use of cellular morphology to assess toxicity of materials is faced with many difficulties. Both physical properties of material surfaces and leachable toxic substances strongly influence morphology. Although the thin film technique is well adapted to morphological studies, many materials and devices can not be tested using that technique. Although standardization and quantitation are difficult, the fact of morphological change under test conditions is significant, and the resolution of interpretative problems should be pursued.

Question to Anyone from J. Lemons (University of Alabama)—The use of the phrase "altered cell morphology" is questioned. Please list and explain the criteria used to quantitatively describe "altered cell morphology." At least 20 different morphologies have been shown thus far, many of which demonstrate alterations from the control tissue-culture cells. It seems that altered cell morphology usually means cell death.

R. Hay—There have been all sorts of debates on that. I think people generally will recognize when cells look healthy and when they do not look healthy. I guess the assigning of a 1 to 4 rating will suffice for many of you. I like the quantitative method that the first afternoon speaker (J. Ulreich) was supporting where you can actually measure what is happening to the cells' response to an inappropriate or unhealthy environment. The quantitation of deoxyribonucleic acid (DNA) synthesis is a very simple thing to do. For $1500 you can buy an extraction unit to measure DNA synthesis with hundreds and hundreds of test wells, literally, in a couple of days. To me that is a lot more valuable than a +1 or a +4 rating.

The replicate units for measuring the amount of labeled DNA in 96- or 24-well plates are coming into use very extensively now. There are half a dozen different manufacturers of those units. At least in our hands with the 96-well microtiter plate you can get very good reproducibility and can run these things off in very little time. What you are really looking at then, is what Judy Ulreich described: the response, not of total cell death, but the cells' replicative function. It is a very sensitive indicator of problems associated with the environment. I believe Dr. Dillingham referred earlier to the fact that the correlation between in vivo and in vitro tests is best when you are measuring that kind of a response of the cells, that is, the replicative function of the cell, rather than whether or not it is alive or dead.

S. Northup—I appreciate your comments about the quantification of the DNA assays and protein assays. What happens is when, in a practical sense, I try to explain what it means when you get 90% protein or 90% DNA. Is that a real number? What does it mean? We had to translate this from our 90% to our 1+ and 2+ system.

J. Ulreich—We are doing the same thing. But we do not have human variability, and at least we have a solid number. Then we can say what is cytotoxic? We have to say we have inhibited the DNA synthesis by 20%. Is this toxic? Shall I throw another wrench into the works? Another problem we run into is if we add a material that might be cytotoxic we might see no incorporation of tritiated thymidine into DNA. However, what we may be doing is causing a simple delay in cell cycle progression; we may have cells that are perfectly healthy but sitting there and not dividing, not synthesizing DNA. Therefore, that is why we always look under the microscope. We can often see that there are no cells remaining when we have a cytolytic response. But we also have to take a look to see if the cells we seeded are there and yet have not incorporated a label. So I think it is an addendum to the method that it is good just to give the cells a quick scan; do not count anything, just take a look so you can get a feeling for what is happening. Thymidine incorporation does not always equal DNA synthesis, unfortunately.

S. Meyers—It seems to me what we are trying to do is predict what is going to happen in a real living system, and it is not on whether the enzyme system compares well to a 1+, 2+, 3+, or whatever system you are working with. The real question is how does it relate back to in vivo reactions. To me that is what it means. If you show 90% inhibition of some kind of an in vivo toxic response, then it has meaning. It does not have meaning because it relates to a 1+.

J. Ulreich—That is a good point. You have to correlate it to in vivo work that is being done.

Question to Kathy Phillips (Extracorporeal) from E. Frisch—Is the method described used to qualify implants or implant materials? If so, would not direct contact be more appropriate in order to also provide a measure of the cytotoxicity potential of the material per se? This cannot be done with extracts.

K. Phillips—There are certain materials that we do test that are used for implants, however, these are few and far between, and usually we receive a "0" score on these. Any materials in our procedures that are used for implants are scored on the same basis but the results are treated differently. The way we score is from 0 to 4+, and we have determined to accept for most components a 0 or 1+; anything above that gets retested. We have a list of components that are used for implants, such as silicone, and if these score +1, it is automatically retested. We do not accept it just on a +1, we will retest it immediately. If it gets a 0 on the retest we will accept that material. If it gets a +1 we will accept it at that point, but it does immediately get sent out for the implant test, and we do go to the vendor to see what is different with this biomaterial that we have received. But if we receive a +2, +3, or +4 on the first test we reject that material. So there is a different criteria for the acceptability of those materials that are used for implants. A direct surface test would probably be better for them but we have never done such a test.

Comment—I would imagine that a certain component of the direct contact test is indeed extraction, maybe a very large component of it.

Question for J. Fehn (Medtronic) from E. Frisch—Are the medical adhesives mentioned used for implants? If so would not any degree of cytotoxicity detected by either overlay or direct contact make the material unsuitable for use as an implant?

J. Fehn—I guess it is a known fact that when you use a medical adhesive you are just tuned to what is in a medical adhesive and are going to get a response there on implant. But if it is something that has been out there in humans that is used by everybody, then it is a known fact that they use them at the time of implant or surgery, or something like that. So just because of the fact that the byproduct of the medical adhesive is acetic acid, if you have not properly cured or done something to that material to drive off all the acetic acid, you are going to get a positive response. There are two medical adhesives for which I can say yes, we get a variation in our results. But what we try to do is to say that we handle it the same way, prepare it the same way, and do everything the same. I cannot explain why we get different results. I think it is a known fact that medical adhesives are toxic and are still used.

E. Frisch—There is actually a difference, of course. The medical adhesive we make does not have an organo-tin catalyst. Most nonmedical grade and industrial type adhesives made for bathtub caulks and miscellaneous things use an organo-tin catalyst to catalyze the reaction of cure. I think that our Medical Adhesive Type A, when fully cured, passes cell-culture testing.

Question to Anyone from M. Brande—Please elaborate on the desirability of using serum-free media if serum contributes to detoxifying a toxicant, which would complicate reading a response.

J. Lontz—Our colleague, serum biochemist, Dr. Richard Holmes, has done quite a bit of work separating cell-growth components in human and various other sera. His work has been published in the literature and indicates also that lipoproteins are inhibitive or highly toxic to cell growth. This leads, or could lead, to a situation of having a balance, as a variable in any sera, of growth factor versus inhibitive factors. Where one dominates over the other for whatever reason, this may be due to the source of some serum processing; the balance could be such that the net growth promotion could override the growth inhibition or conversely could be overridden. There is also the feature of protein complexes that may induce or inhibit growth depending upon the nature of the complex or its decomplexed form. For this reason, a multicomponent medium like Holmes A-3 with some 160 chemically defined components serves to augment dialyzed, and so detoxified, serum from the adventitious toxicant for species-identical cell-serum systems. There is increasing awareness with new media continually being published, that specific cofactors, some from chemically defined medium and some by separations from sera or plasma, contribute to highly specific growth responses and some to highly inhibitive responses. There is a complexity of serum species versus cell species. As a case in point Harper (*Cell Biology International Reports,* Vol. 6, 1982, pp. 163–169) singles out the cofactor retinoic acid as either a growth stimulator or growth promotor depending upon the combination of cell species versus serum species.

J. Ulreich—Certainly in our experience the presence of serum does reduce the toxicity of some materials because some components such as albumin bind a lot of substances. However, tissue culture is an attempt to reproduce the in vivo situation in which tissues are continually bathed if not in serum at least in plasma. Therefore, I think that probably, as long as you are consistent in your use of method, either serum free or containing serum, each method has certain advantages.

S. Brown—John Lontz mentioned problems of identifying different factors in serum. Some of the other speakers mentioned that they had used fetal calf serum and "because the price went up" they changed to newborn. I am wondering if any of the others feel that there may be some changes in the sensitivity of the methods or the results because of the change in the type of serum. In other words, has anyone else had the sort of problems that Lontz was looking at in terms of serum contaminates and special growth factors, and so forth.

E. Dillingham—There is no question that the quality of serum varies and that controls must be run on each lot to assure reproducibility in testing. We have used fetal calf, newborn calf, and horse serums in our tests and with appropriate controls reproducibility can be achieved. However, the simple specification of the type of serum is not adequate for standardization of test results among laboratories. The quality of serums used in different laboratories varies as does the physiological properties of the cell lines used in different laboratories. We are going to have to face the fact that there is an intrinsic variability in in vitro tests resulting from those factors and the methods of quantitation used, which are often only semi-quantitative. The inhibition of cell-growth test has a 29% confidence limit based on the results of different technicians over long periods of time. This includes the variability of extracts tested, which we have discussed. The idea that a well defined serum per se will significantly improve the reproducibility of in vitro tests is a little optimistic.

J. Fehn—I would be interested in what Dr. Hay has to say since you are developing a lot of these cell lines. Which is the type of serum that you find best?

R. Hay—I do not like to endorse any one supplier, but we do use several suppliers, and we have obtained good quality serum. We have used serum from several companies in the last five years or so, and we have had very few problems. We do test all of our serum on cells after we purchase them, for bovine viral diarrhea (BVD) virus, growth properties, and so on. The sorts of tests that we generally subscribe to are available in the quality control pamphlet.

S. Meyers—Given the variability of the serum and variability from lab to lab it seems to me that if you used your controls properly, that is, negative and positive controls, you should be able to tell from time to time and lab to lab if the material is highly toxic or mildly toxic or nontoxic.

Question for D. Magoffin from S. Brown—Would you explain your comment concerning "type of toxicity?"

D. Magoffin—I was referring to something that was also demonstrated in J. Ulreich's talk this afternoon. That is, when you are measuring various biological end points, it is possible for a toxin to affect one cellular process to a large extent and then have very little effect on another cellular process. A second toxin could affect only biological end points of a totally different nature. For instance, protein synthesis could be affected by particular classes of toxins, whereas other classes of toxins may not affect protein synthesis directly, but they may cause breakdown of the cellular membranes. The literature is full of examples of compounds that affect specific cellular processes. These effects are frequently the basis of bioassays.

J. Ulreich—This might be an important deterrant to creating a set of standards, although I support the establishment of a set of standards. I would love to be able to take a standard and use it in my system and say, "OK, this is my sensitivity, and we can compare it to so and so's." It might relieve me of the responsibility of testing some other materials. But the question is if you have a standard that affects the cell membrane, how do you compare it to a test substance that inhibits DNA polymerase? There is a problem because we are comparing apples and oranges. I fully support the idea of having a standard, but this might be a good reason to have more than one standard that could then be used in different dilutions under different circumstances.

T. Walker (CSA)—It sounds like we might be looking at two or three different control materials, positive controls as well, each having a different effect on cell function and cell viability.

R. Schlesinger—I do not know whether Judy's (Ulreich) uptake procedure could be used by all of us, but something that measures a physiological function, rather than a morphological end point, or death, certainly is more informative. As she indicated in her answer to the question that was asked by Denis (Magoffin), if a given toxin has a physiological effect, measuring the effects on that physiological end point will be a much more sensitive assay.

Question to L. Stroempl from R. Hay—You mentioned that the toxic end point makes your test quantitative. Is this not linked with the test time? That is, can you not extend or restrict the end point by varying the test time? In the same vein can you really be sure of no toxicity, (that is, tests on others) with in vivo comparisons?

L. Stroempl—To the first part of the question, yes, cell-contact holding time can have an effect on the titer in a series of dilutions. In fact, when we offer that titration test to an investigator for his material, we will offer it for a period of up to 72 h comparable to what we do as a routine cytotoxicity test. On the other hand when we use it in a week to week control setting, we have terminated the results and defined the titer of control values at 24 h. Now I do not understand the second part of the question.

T. Walker—I would read it as a question of whether any cytotoxicity assay really assures you of no toxic effects.

L. Stroempl—No, it probably does not. I am not sure why I got asked that question.

R. Hay—I think you said specifically that you categorize your compounds as nontoxic, intermediate, or toxic.

L. Stroempl—We would not consider our end category to have absolutely no toxic effect (but I am sure that I said it in that way). Obviously what we are saying is no observable toxic effect.

R. Hay—I would imagine that it is very hard to rule out toxicity, and that is why I was concerned about control and how people, for example, that are giving blood transfusions or that are using syringes and needles, can really be sure that they are not introducing something that over the long term might be very detrimental. How can recipients be assured that the stuff being put in their veins is not going to affect them ten years later? I guess the tests really are not that sensitive. The C3H 10 T ½ test that I referred to is a very sensitive test, but perhaps for human use it is not a good one to recommend. For commercial organizations I think it might be too difficult to apply on a wide scale. It is just too time consumming.

L. Stroempl—We have to remember too, that cytotoxicity testing per se is not essentially evidence and not the be all and end all in biocompatibility testing. If we are to ask questions about long-term effects, these can probably be better answered with other methods than a cytotoxicity test. I do not think it is a drawback to cytotoxicity testing that it cannot answer every question about the safety of a material.

S. Brown—It is really only the first step in the screening of new material, or a method that can be used for looking at variation in a manufacturing process in a quality control sense.

Comment—In any biological system that you are talking about, sterility or any of the other tests that device manufacturers use, there is always some risk. You can never assure the patient that with an "absolute sterility;" you cannot guarantee that he will not get endocarditis from our heart valve. We can only give him a probability. In any of our tests there is always some patient risk.

S. Brown—There is also the risk-benefit ratio associated with all implants: how much good are you going to do for the patient by putting the thing in as opposed to how much risk you will take by not having it in there?

Question to R. Horres from R. Hay—In your indication that cells are not used beyond 24 passages from origin, does this mean beyond 24 population doublings (PDL) total, beyond 24 passages from a specified seed stock, or beyond 17 PDL + a number of passages with PDLs unknown?

R. Hay—I believe that question was answered earlier. The only reason for posing it was to once again emphasize the importance of PDLs as opposed to passages because passages basically mean very little when one is dealing with a diploid cell strain.

Question to K. Phillips from R. Hay—Since your products are in contact with blood, have you compared results on WI-38 with those on primary human leukocytes over the short term, or with lymphocytic or endothelial lines?

K. Phillips—No, nothing except WI-38 or MRC-5.

J. Ulreich—We have done some work with zinc that is peripherally relevant. Zinc does inhibit the functions of polymosphonuclear leukocytes (PMNs) as well as fibroblasts and also macrophages. We are not saying there is an equivalence; it was asked whether blood cells had the same response, and in the case of zinc toxicity the answer is yes.

Question for E. Dillingham from R. Hay—Since tests for cell growth in vitro appear to correlate best with in vivo toxicity, would you recommend tritiated thymodine labeling tests as opposed to other isotope labeling tests for quick materials toxicity testing?

E. Dillingham—I am sure that the next round of test development will include radiochemical procedures. We have used radioactive amino acids to assess inhibition of cell growth, particularly for the testing of specimens that contain protein, which interferes with the routine colorimetric assay of total cellular protein. This combined with pulse-labeling with thymodine at intervals over the growth cycle or assay period gives considerably more information on the toxic properties of the test solution or extract than the simple 72-h assay. The extension of the routine data base to include such parameters is ideal, but we are not talking about simple, routine tests to accomplish this. It is the direction of the future, however, and, I think, the right direction.

Question to E. Dillingham from R. Wallin—What conclusions do you reach concerning the value of cell-culture methods in testing device components? What is their relative value and with what in vivo test do they agree best? Secondly, why do you feel polyethylene glycol 400 is still important as an extraction medium if oil and water are included?

E. Dillingham—Polyethylene glycol is an intermediate polarity extraction medium. The polyethylene glycol extract tests are correlated with cottonseed oil extracts, and both are about equal in correlation with tests directly on the material (agar overlay and implant tests). The important thing to remember is that the use of only one of the extracts would miss the positive response in a significant percentage of the tests since the univariate correlations are in the range of 0.6 to 0.8. A positive response (lethality) in the systemic toxicity test of the polyethylene glycol extract is rare but is frequent enough to justify the retention of both the systemic toxicity test and the polyethylene glycol extract in the screen.

I would rank the inhibition cell growth as the most sensitive and best correlated with in vivo tests based on multiple regression analysis against all primary screen tests. The hemolysis is also well correlated with in vivo tests but is a less sensitive test. Approximately an order of magnitude higher concentration of a leachable or toxicant is required to induce hemolysis than is required to induce response in tissue culture. With respect to extraction media, the profile of response with the different extraction media gives an insight into the leachability toxicity of the components leaching, although these parameters are confounded in all extract tests. Aqueous and nonpolar extracts tend to be negatively correlated, and the histopathologic response shows clear differences between those extracts. The tissue-culture agar overlay test is the most versatile test, since diverse materials and extracts can be tested. The inhibition of cell-growth test is the most sensitive; the hemolysis test the most simple and convenient.

Question to L. Smith and K. Phillips from R. Hay—Do you restrict the population doubling range of your test fibroblast cultures or do you notice any variability when cultures at widely differing PDL levels are used or both?

L. Smith—No, we do not restrict the numbers that we use in the tests. What we normally do rather than make counts on the cells is to let them grow, once they are seeded in a test well, until they reach what we consider a confluent monolayer, whether it be in two days, three days, or four days. We determine microscopically when they are ready for the test cycle. So we choose to monitor cell population that way rather than to make counts and add a calculated number of cells to the well.

K. Phillips—That is basically the same as we do. When I first started doing tissue culture, we used to count them and add a certain number of cells to each, and it turned out that we were diluting it the same each week I made plates. Now I just basically make it the same and during the following week whenever my plates are ready, that is when I do the test.

L. Stroempl—Although we also do not really follow a whole population doubling per se, our weekly titration data have shown that from the time that we originally purchase a new culture of the cells until the time when they have reached basically the end of their useful life in our lab, we are seeing no trend, whatsoever, in terms of their sensitivity to a control toxin.

Second Part of the Same Question—Is your test strictly based on morphologic examination or have you tried quantitative tests such as incorporation of tritiated thymodine or quantitation of cell proliferation?

L. Smith—No, it is strictly microscopic evaluation. As I said before, although we have in the past used a 1 to 5 gradation system, at present we rate solely on a pass/fail basis, cytopathic effect or no cytopathic effect. Any alteration in morphology, whether it is just the cells look "sick" or if they are dead as doornails, we consider both reactions cytopathic effects.

K. Phillips—We only use the microscopic.

The Third Part of the Question—Do you use any solubilizing agent, such as dimethyl sulfoxide (DMSO) to deliver potential hydrophobic chemicals on your test material to the test cells?

L. Smith—No, I cannot think of any examples where we have encountered that type of situation.

K. Phillips—No.

Question to Anyone from S. Brown—Addressing the issue of end use application, which is the basis of the HIMA/PMA documents and the basis of ASTM Practice for Selecting Generic Biological Test Methods for Materials and Devices (F 748); do you feel we should focus on the issue of cell line or on the method of test in terms of when we try to look at blood contact, long-term, short-term, and so forth? I think basically at this stage we are not going to try to address that issue just to try to get something down on paper. Anyone want to make a comment to that? (No response.) Another question, and this would be addressed to someone using direct contact or agar overlay: if you take a slice of tubing, do you ever look at the response on the inside of the tubing, as opposed to the ring you get on the outside?

E. Dillingham—You can do this by selectively cutting and exposing surfaces. We have done this with respect to tubing; it generally depends on whether you are looking at the total tubing or not. If you have a pure polymer tubing you see little difference with respect to the surfaces exposed. If you have a coated tubing, you may see a little, but most coatings do not inhibit leachability, so there is not really much difference with respect to surfaces exposed.

L. Veresh—In our testing, rather than using it as a quality assurance test, we are using it as a screening test for new materials that are being evaluated that are being considered as extrusion grade materials for making medical grade tubing materials. So when we get a sample of that kind of tubing material in to test, yes, we cut it crosswise, lengthwise, any way you can so you are looking at as many different surfaces the outside, inner, cross-sectional surfaces, to see if there is any difference. We have not seen any so far. But we keep looking because there is always that possibility.

J. Lontz—We had trouble. We used Teflon tubing, the same bore as one of the latex tubing, and the problem is there was no immobilization fixation for some reason, I do not know why. In handling we go from a decontamination room to an objects room. Our problem was that there was a sliding across petri dishes and that sluffed things off pretty badly so we could not find it. Does anyone have any way of immobilizing rubber latex at the bottom of a glass?

L. Veresh—Just weight it down with the negative control.

J. Lontz—We use Teflon as a negative control. If it gets too heavy it slides. If you take a 2 to 4-mm pipe of Teflon tubing, one the company uses for arterial access, our big problem is that they slide. We make sure the diameter is larger than the height so it will not tip over. So our problem has been the sliding. We wanted to look at that examination to see what is inside, if there is access to the cell, and if there was blood-serum problems. So if anyone knows a way of immobilizing I would be very happy to hear it.

S. Brown—I was thinking even with your samples with the little holes in the middle if you get something analogous to a surface energy phenomenon inside as opposed to outside. Does that make any difference?

M. Brande—We have developed a three-arm plastic insert that goes into a 60-mm well and in that well we put our test material and a glass disk that holds the material in the center of the well so it does not slide around. The adapter keeps the test material held.

Question to Anyone from R. Yapp (*Carbomedics*)—With the advent of porous materials designed specifically for tissue in-growth fixation, can you suggest test methods which: (1) Can demonstrate a potential metal or polymer toxicity caused by increased implant surface area in an otherwise nontoxic material? (2) Can be used to determine cleanliness and sterility of such materials? and (3) How often should such testing be done to guarantee a safe product?

S. Brown—I would guess the first part of it is simply that it is a surface area phenomenon and I think we have addressed certain aspects of the surface area to volume ratio. I would not think the phenomena would be any different; you would want to use more extract liquid if you are doing an extraction procedure. I think that is probably not a major issue. I think the other question is not an easy one to answer.

E. Frisch—We did cell-culture quality assurance on Proplast®, by the same direct contact method used with solid silicone elastomer. If a material does not biodegrade and is inherently noncytotoxic, such as with Proplast, then neither the quantity of the material tested nor the surface area should change the test result. We addressed the sterilization question because the material was furnished nonsterile with recommendation of sterilization by steam autoclave. With Proplast the material has heat resistance that far exceeds the steam autoclave sterilization temperature, and, in fact, is exposed to significantly higher temperatures during its manufacturing.

S. Brown—Ladies and gentlemen, we have come to the end of our written questions and to the end of our allotted time. On behalf of the organizers (myself, Tom Nickel, and Tom Walker) and our co-workers in ASTM F4.2.6 on Biocompatibility, I would like to thank you all for your time and contributions. From my point of view, this day has been a great success. You have presented us with the factual information we needed to come to grips with the problem of writing cell-culture test method standards, and you have provided us with guidance as to the type and specificity of what I anticipate will be usable benchmark test methods. Again, I thank you one and all, and call this session adjourned.

Summary

The papers included in this volume are organized in groups according to the nature of the papers. The first few papers discuss specific methods with specific cell lines. The paper by Smith discusses the direct contact method with human WI-38 cells that has been used for more than ten years by Dow Corning. In the paper the author discusses the extensive use of this test for new materials screening, as well as some studies they have done to verify the equivalency of the WI-38 cell line with several other human-cell lines that can be used as backups. As a demonstration of the sensitivity of the method, data are presented on the cytopathic effect of different ethylene levels of ethylene oxide residual from gas sterilization. The second paper by Ettorre and Philips presents a minimal essential medium (MEM) elution method that also uses the WI-38 cell line. This paper is presented as a protocol and gives a grading scheme and complete list of all the materials necessary for conducting the test. This paper is actually the starting point for the ASTM Committee F-4 development of a MEM elution test method. The third paper by Fehn and Schottler discusses two methods of agar overlay and MEM elution, but in this instance using the L-929 mouse fibroblast cell line. In this they discuss the differences between the two methods and when one would be selected over the other. The paper by Johnson and Northup discusses the use of L-929 cells in a battery of fluid and agar overlay methods. The details of the methods and their rationales for selection of one method over another are presented, based on the nature of the material. In addition, the results of some in vitro versus in vivo screening tests are presented. The fifth paper by Horres et al uses WI-38 cells in cytotoxicity and cell-growth inhibition tests. In this paper the authors discuss the differences between cytotoxicity and growth inhibition, both in terms of methodology as well as application. In addition, they discuss the question of types of media to be used for extraction of plastics and report that extraction in culture media itself is better than other extraction fluids.

Two papers are presented that address themselves to the question of in vivo and in vitro correlation. The paper by Dillingham goes beyond cell-culture testing and addresses a battery of tests that are used at the University of Tennessee as a primary acute toxicity screen. Dillingham discusses the rationale for selecting the 14 tests that are used as part of the toxicity screen and the rating scheme, which they refer to as the "Cummulative Toxicity Index." Tests are performed on materials as well as extracts in water, saline,

PEG-400, and cottonseed oil. The tests used are agar overlay cell culture, intramuscular implant, hemolysis, intracutaneous injection, systemic toxicity, and cell-growth inhibition. These tests include three in vitro and three in vivo assays for biologic reaction to extractable moieties. A very detailed, statistical correlation between in vitro and in vivo performances is discussed. Their analysis includes the results from extensive interlaboratory comparisons of results of testing on different materials. The paper by Genova et al addresses the issues of levels of toxicity, vis-a-vis end use application, as well as the question of in vivo and in vitro correlation. In order to pursue these questions, they have developed in-house reference positive and negative standard materials and demonstrate correlations between intramuscular and agar overlay responses. In addressing the question of sensitivity of cell-culture methods they make a comparison with what they refer to as the minimal animal response equivalent (MARE) and conclude that anything with a response greater than MARE in vitro would be expected to be toxic in vivo. The material presented in these two papers represents a significant contribution to the utilization of in vitro results for predicting in vivo performance.

The third group of papers is concerned with the design of cell-culture methods to closely mimic the intended in vivo application. The paper by Lontz et al discusses a specific cell-culture method, using human excised donor cells and specially prepared growth media that has been developed for the specific application of screening materials for oro-facial implants and prosthetics. This paper represents work directed toward tailoring an in vitro toxicity assay as closely as possible to the end use application of the material being screened. The paper by Meyers et al also addresses the question of using specific cell lines and a method that would be appropriate for a specific end use application, in this case blood contact. They discuss the literature and the history of cell-growth and cell-culture methods and bring out some of the questions concerning standardization and application of in vitro methods for new material screening.

The next two papers discuss the use of radioactive labels for quantitative cell-culture screening. The paper by Magoffin uses L-929 cells and ^{51}Cr release. Utilizing chrome release and several dilutions of phenol, the authors conduct a series of tests to compare the sensitivity of ^{51}Cr release agar overlay and growth inhibition test methods. In addition, they compare these in vitro results with the response of rabbits to intracutaneous injection of phenol in different concentrations. Their results demonstrate a similar type of response of three culture methods, and a correlation between the ^{51}Cr release in vitro and the response of the rabbit in vivo to comparable levels of phenol. In contrast, the paper by Ulreich and Chvapil discusses methods other than ^{51}Cr release for radioactive tracer assessment of various cell functions. In a series of assays using 3T3 cells, tritiated thymidine was utilized to assay for cell viability; tritiated tryptophane is used to assay for synthesis of noncollagenous protein, and $^{35}SO_4$ was used to assay for the synthesis of glucosaminoglycan.

They present results using collagen sponges treated by different methods with different levels of extractables, to demonstrate the sensitivity and selectivity of the different methods for cytotoxicity testing.

The final paper by Hay is from the American Type Culture Collection (ATCC). In this the methods for cell banking, quality control, and cell line development and characterization are discussed. The author also discusses some of the characteristics of the different cell lines, indicating the appropriateness of some cells for certain applications. The paper is illustrated with examples of the characteristics that are examined during quality control tests of cells supplied by ATCC.

The workshop discussion was organized into four parts. The first three parts addressed questions that were raised from the Chair, yet throughout the discussion there was a repeated return to Question 1. Question 1 was in essence, "What should ASTM Committee F-4 be doing about cytotoxicity test methods? Should we write standards; should they be generic or specific; who would use them?" In this discussion the consensus was that we should write specific, benchmark type methods that could be used for interlaboratory comparison. Question 2 related to the utility of developing standard reference charts for ranking cell reactions. The consensus was that this would not be effective. Part 3 addressed the problem of the development of reference materials as positive and negative controls for toxicity testing. With regard to the positive control, two possible solutions to the problem were considered: one was a specific material containing a toxin, and the second was utilization of a liquid that could be titrated to a desired level of toxicity. It was clear from this discussion that there is need for reference materials that can be used routinely as well as to check the sensitivity of tests in interlaboratory comparisons. Part 4 of the workshop consisted of reading written questions from the audience that had been submitted during the morning session. For the most part, these questions were addressed specifically to authors to clarify points that they had made. This open discussion format was used so that questions could be answered by the authors as well as be discussed by the audience in general. These discussions have been reviewed and edited by the speakers. In some instances the text has been changed from what was said, but the substance and meaning is unchanged.

Index